WITCHES, SCIENTISTS, PHILOSOPHERS: ESSAYS AND LECTURES

THE WESTERN ONTARIO SERIES IN PHILOSOPHY OF SCIENCE

A SERIES OF BOOKS
IN PHILOSOPHY OF SCIENCE, METHODOLOGY, EPISTEMOLOGY,
LOGIC, HISTORY OF SCIENCE, AND RELATED FIELDS

VOLUME 65

ROBERT E. BUTTS

WITCHES, SCIENTISTS, PHILOSOPHERS: ESSAYS AND LECTURES

Edited by

GRAHAM SOLOMON
Wilfrid Laurier University,
Waterloo, Ontario, Canada

KLUWER ACADEMIC PUBLISHERS
DORDRECHT / BOSTON / LONDON

A C.I.P. Catalogue record for this book is available from the Library of Congress.

ISBN 0 7923-6608-5

Published by Kluwer Academic Publishers,
P.O. Box 17, 3300 AA Dordrecht, The Netherlands.

Sold and distributed in North, Central and South America
by Kluwer Academic Publishers,
101 Philip Drive, Norwell, MA 02061, U.S.A.

In all other countries, sold and distributed
by Kluwer Academic Publishers,
P.O. Box 322, 3300 AH Dordrecht, The Netherlands.

Printed on acid-free paper

Printed in the Netherlands.

Table of Contents

PREFACE

Robert Butts (1928 - 1997) was a well known and respected historian and philosopher of science. He played an important role in establishing History and Philosophy of Science as a major field of study in Canada and the United States. He established and gave direction to the internationally respected research profile in HPS studies at the University of Western Ontario. He was a two term editor of *Philosophy of Science*, a leading journal in the field, and also was a long serving Managing Editor of the *Western Ontario Series in the Philosophy of Science*. He made significant contributions to scholarship on Kant and Leibniz, and was among the first to rehabilitate the work and reputation of Whewell and other 19th century philosophers of science.

After his retirement he continued to teach, offering an undergraduate course called "The Metaphysics and Epistemology of Witchcraft", in which he applied many of the important lessons he'd learned in his HPS work. The course was proposed and developed in 1978 by Thomas Lennon and became very popular, attracting some 400 students per year when Butts taught it. General interest in the topic and Butts's reputation as a scholar and lecturer led to an invitation to deliver a series of three lectures as part of the prestigious Tamblyn Lectures series at the University of Western Ontario. He gave the lectures over three evenings in March 1996.

Butts died in 1997, before he could make a decision about revising and publishing the Tamblyn Lectures. Members of the audience and various colleagues and students hoped the lectures would be published in some form or other. There is a good deal of popular interest in the history of the witchcraft persecutions, and the insights and perspective of a historian and philosopher of science deepen and clarify the issues.

This collection of Butts's lectures and essays brings together the Tamblyn Lectures and a related lecture on witchcraft, and eleven papers on historically significant philosophers and on issues in the history and philosophy of science. Butts saw himself as a pragmatic realist, combining what he took to be the best aspects of logical empiricism with a historically informed neo-Kantian pragmatism deeply appreciative of the methods of science. His work on witchcraft is thoroughly informed by his expert grasp of basic issues in the philosophy of science and his respect for the place of science in culture. The papers on Leibniz, Hume, Husserl, and Kant, and on issues in history and philosophy of science, while more specialized than the lectures on witchcraft, are accessible and understandable (and of much interest on their own), and supply useful context for the witchcraft lectures. We see in these papers on philosophy and science that, for Butts, scientific rationality is the major defence we have against the various conditions which encourage witch hunts and all the pain and terror that go with them. We see behind all the articles a fine philosopher at work, trying to locate the distinction between the rational and the irrational, trying to describe a kind of rationality useful, perhaps essential, in the struggle to avoid barbarism and preserve humane values.

Chapters 1 and 2 are lectures on witchcraft; Chapter 9 is a lecture on Kant on illusion; and Chapter 13 is a lecture on the history of scientific philosophy from the founding of the journal *Erkenntnis* to the publication of Kuhn's *Structure of Scientific Revolutions*. Chapter 1 is the unrevised text of an invited lecture given to the Annual Conference of the Seminar for Medieval and Renaissance Studies, at the University of Western Ontario in 1992, and is a useful introduction to the Tamblyn Lectures, concisely expressing some of the philosophical interest he found in the topic of witchcraft. Chapter 2 is the Tamblyn Lectures. Some bibliographical footnotes have been added; otherwise the text is as it was delivered. Chapter 9 is a draft of a paper he had nearly completed and was circulating among friends and colleagues. Earlier versions of it had been given as invited lectures at the Pacific Division meeting of the American Philosophical Association in 1994, and at the University of Rochester in 1996. Chapter 13 is the fourth and last draft he was able to complete of a paper which was posthumously published in *Experience, Reality, and Scientific Explanation: Essays in Honor of Merrilee and Wesley Salmon*, edited by Maria Carla Galavotti and Alessandro Pagnini (Dordrecht: Kluwer, 1999), pp. 1-14. It is reprinted here with the kind permission of Kluwer Academic Publishers.

I thank Valentine Rodger for permission to publish these uncompleted drafts by her husband and for permission to collect and publish the following previously published articles. I thank William Demopoulos for advice and Nathan Colborne for assistance. The remaining acknowledgements are as follows:

"Leibniz' Monads: A Heritage of Gnosticism and a Source of Rational Science", *Canadian Journal of Philosophy* 10 (1980): 47-62, published by The University of Calgary Press, reproduced here by the kind permission of The University of Calgary Press.

"Rationalism in Modern Science: d'Alembert and the 'esprit simpliste'", *Bucknell Review* 8 (1959): 127-139, published by Bucknell University Press, reproduced here by the kind permission of Bucknell University Press.

"Hume's Scepticism", *Journal of the History of Ideas* 20 (1959): 413-419, Journal of the History of Ideas, Inc. Reprinted by the kind permission of the Johns Hopkins University Press.

"Husserl's Critique of Hume's Notion of Distinctions of Reason", *Philosophy and Phenomenological Research* 20 (1959): 213-221, published by the International Phenomenological Society, reproduced here by the kind permission of the Editor of *Philosophy and Phenomenological Research*.

"The Role of Arational Factors in Interpretive History: The Case of Kant and ESP", in J. R. Brown (ed.), *Scientific Rationality: the Sociological Turn* (Dordrecht: Reidel), pp. 209-247, reproduced here by the kind permission of Kluwer Academic Publishers.

"Kant's Theory of Musical Sound: An early Exercise in Cognitive Science", *Dialogue* 32 (1993): 3-24, published by Wilfrid Laurier University Press for the Canadian Philosophical Association, reproduced here by the kind permission of the Editor of *Dialogue*.

"The Hypothetico-Deductive Model of Scientific Theories: A Sympathetic Disclaimer", in William R. Shea (ed.), *Basic Issues In The Philosophy Of Science* (New York: Science History Publications, 1976), pp. 36-57. In Public Domain.

"Methodology and the Functional Identity of Science and Philosophy", in J. Hintikka, D. Gruender, and E. Agazzi (eds), *Probabilistic Thinking, Thermodynamics, And The Interaction Of The History And Philosophy Of Science (Pisa Conference Proceedings)*, Vol. II (Dordrecht: Reidel, 1981), pp. 253-270, reproduced here by the kind permission of Kluwer Academic Publishers.

"Sciences and Psuedosciences: An Attempt at a New Form of Demarcation", by Robert E. Butts, from *Philosophical Problems of the Internal and External Worlds*, John Earman, Allen I. Janis, Gerald J. Massey, and Nicholas Rescher, Eds., © 1993 by University of Pittsburgh Press. Reprinted by permission of the University of Pittsburgh Press.

Graham Solomon
Waterloo

PART I

WITCHES AND WITCH-HUNTERS

CHAPTER 1

TWO STORIES ABOUT EVIL: CHRISTIANITY AND THE CREATION OF WITCHES[1]

Evil is something personal: it happens to *me*. That you suffer evil is something I can only know by analogical inference or by empathy. But we all know what evil is. It is not a philosopher's abstraction, it is real, it is *felt*. Feeling is of importance here: only creatures capable of experiencing *pain* can suffer evil. I do not here refer to painful pinpricks, the disappointment of unrecognized achievements, or the sorrows of a lost love. What I have in mind is an evil presence, an immediacy of pain so sharply defined as to be seemingly inexplicable. In Dostoyevsky's *The Brothers Karamazov* Ivan says to Alyosha:

> Imagine a trembling mother with her baby in her arms, a circle of invading Turks around her. They've planned a diversion: they pet the baby, laugh to make it laugh. They succeed, the baby laughs. At that moment a Turk points a pistol four inches from the baby's face. The baby laughs with glee, holds out its little hands to the pistol, and he pulls the trigger in the baby's face and blows out its brains.[2]

Examples abound: how could Sophie's choice possibly be a *choice*, offered as it is with no possible outcome but the spiritual death of a mother? US GIs in Viet Nam murdering innocent civilians, to what purpose? Countless mothers baking countless blueberry pies eaten by countless sons, some of whom go to war and are transformed into monsters. What is the connection? One of my beloved teachers, a gentle soul who taught symbolic logic in the basements of burned-out buildings in Warsaw during the Second World War, put this challenge to me: A Nazi officer condemns 12 innocent civilians to death, stands silent in the face of the victims while his soldiers murder them with rifle fire; and then retires to his rooms, puts on a record of some sublime classical music on his turntable, and sits listening, quiet, relaxed and content. Explain to me the connection between these events and you will have solved the problem of evil.

Fortunately for me, I have not been invited to come here to solve the problem of evil. Against the minds of Aquinas and Leibniz, Dostoyevsky and Kafka, my own poor spirit retreats in a mood of endless questioning. I have a

---

1 [Previously unpublished]

2 Translated by Constance Garnett (New York: Vintage, 1955), p. 238.

hunch, however, that Sartre was right to think that Ivan's identification of the fostering of evil with the devil *in us*, is a key to the solution of the problem. We humans seem to have an inexhaustible capacity to want to turn other persons into *things*, *objects*. Hell, said Sartre, is other people.

Witches. We are here to learn something about witches, and witches – this is my theme – have much to do with how we understand evil. Witches are *victims*, members of a target group whose parameters are defined in canon or civil law or in the mood swings of a society. But victims, in this or that portion of history, are interchangeable. The witches of the 16th and 17th centuries, the witches of Salem, the pre-Reformation dissidents in Albi, the Jews throughout the history of modern western 'civilization', the persons named on McCarthy's never-disclosed 'lists', those next to suffer from allegations of 'political incorrectness' or lack of an 'authentic voice' – these are all, as I understand the connection between enforcement and victimization, interchangeable social 'roles'. The religious offense is heresy. In disquiet moments of history, many of us become secular heretics.

I have presented some presuppositions, some intimations of the context in which I approach the attempt to understand the evil we put upon one another. Down to business. The title of my paper promises the telling of two stories about evil. Both stories evade the environment of Dostoyevsky's setting of the problem *– both stories assiduously eliminate the possibility that humans may have produced the devil in their own image, for both stories have much to do with the role of God in a creation that is frequently beset by evil doings.*

Story 1: the theology of evil. If God is a fundamental factor in the equation, it seems that there are only three (there will be variations and irrelevancies) possible solutions to the problem of evil. One can be eliminated at once. It is the idea that the universe we know was created by two beings--there are two sources of the universe: one creating that which is good (God) and one creating that which is evil (Satan). Only a moment's reflection will result in one's concluding that this account will not do. The proposed solution accounts for the existence of evil, but at the enormous cost of jeopardizing the unity and the perfection of God, the *goodness* in the creation that is supposed to be expressive of the nature of the creator. Cosmic dualism, given the Judeo-Christian commitment to monotheism, must be declared heretical.

The theology of evil – in Christian dress – seeks to correct and subvert this dualist theology. The thesis is simple: God is super-abundant goodness, so limitlessly perfect that we poor humans can only have an analogical understanding of this being whose very meaning is to be. God is super-everything: omnipotent, omniscient, prudential. Such a being cannot generate evil. Yet evil exists; it exists as *privation*, as *lack*, as *absence*. God, of course, cannot create absences, only presences. So the theological story lets Him off the hook.[3] There is evil, but He

---

[3] I realize that present winds of doctrine dictate the use of 'Her' in this context. But both of the stories I am reviewing are stories about a male God – Christianity has always been a patriarchal story, in whatever version it is told – as we all know. Personally, I have no stake in God's gender. The

cannot initiate it. God is not responsible for my lack of will to act in ways that are always good. God is not responsible for the action of the Turk who shot the baby, nor for the spiritual degeneration of the eaters of blueberry pie. God simply engenders a series of beings ranging from Himself as most perfect, through the angelic realm, to humans, to other animals, to brute matter. The Great Chain of Being, all of it in some respect perfect, no positive evils.

It is important to pause to reflect on the role played by angels in this chain, for we will need angels aplenty when we move on to the second story. The theological story requires the postulation of angels on the basis of cold logic (indeed, everything in this story is to be told on the basis of cold logic). If there are no angels, the gap in the Great Chain of Being between God and humans is too great. God knows everything on the basis of one idea only: His idea of Himself. Humans know next to nothing but have a huge supply of ideas. The intermediaries – the angels – share with God that they know everything (they never have to remember because they never forget), and with humans that they need lots of ideas in order to achieve this perfect knowledge. Angels, on this scenario, play a role as epistemological missing links.

Evil as privation and the Great Chain of Being. A story rigorous in logic and light in reassurance – a story that would satisfy the seminarians and a succession of Popes right down to the present occupant of the Vatican chambers, but a story that would disappear in a cloud of befuddlement when sounded from the pulpit. Yet too a story that was on hand when Innocent VIII issued the Bull in 1484 that would be the clarion call to arms against the witches. But alas a story that could not, by itself, have provided the conceptual foundations for witch hunts. No alleged witch knew the story, very few of the accusers knew it either. It solved the problem of evil, but it was completely out of touch with the realities of pain and suffering. Those in positions of dominance could rest content that they knew the truth about evil, those in positions of dependence actually felt evil. For the dependent, acquiescence could only be achieved by provision of a quite different story about evil – Christianity had one ready for the occasion.

Story 2: the personification of evil. In the Middle Ages and the Renaissance there existed persons identified as witches. We are familiar with the personality profile: a typical witch is a woman, lives a poor life, normally in a small village, is a social dissident, is over 35 years of age, is responsible for all sorts of evil doings, is non-literate.[4] We can of course identify such persons in all ages of history. What makes our period unique is that the witches are in certain ways connected with devils. In this story, evil takes on a personality; brought down from the chilly heights of pure logic, evil regains its place as human suffering. It is the

---

hermaphroditic Monas of the hermeticists has as much appeal as most

[4] It is said that in the 16th Century the average *male* urbanite received as new information, over a lifetime, about what is contained in a single daily edition of *The New York Times*. It is important to remind ourselves of the fact that in the period we are concerned with, reliable information was about as scarce as good plumbing.

pact with the devil that permits us to assume that our ills are brought about by an instrument of the devil: the witch.

Evil personified, evil made personal. Every witch hunter, Catholic or Protestant, cited *Exodus XXII, 17*: 'Thou shalt not suffer a witch (sorcerer, poisoner) to live'. To be brief: execute those who practice maleficent magic. For the ancient Jew, this proscription had nothing to do with belief in the actual existence of witches; it was intended to abolish a particularly awful form of idolatry, and to protect reverence for the unity of God.[5] Christianity had its own gloss on this text. The witch must die, because the witch's spirit is contaminated with the will of a devil. To execute a witch is to prevent further evil doings by that witch, it is also to do her the favor – by burning her contaminated body – of readying her soul for repentance and redemption. How subtle are the changes in logic as we move from story to story.

Without devils, no witches. But also: without angels, no devils. The second story about evil requires that we replace the angel as abstract missing link with the angel as real being. The telling of the story requires a knowledge of Christian angelology, a mythology of great power, one that could be understood when sounded from the pulpit, one that would vitally nerve the witch hunts in ways unavailable to theology. The theology, it was thought, was the *Bible* made rational, revelation rendered logical. It mattered little that the second story had very little literal Biblical endorsement: the story put the pain of evil where it belongs – in human suffering. Anyway, it would be a long time before accused witches would find it possible to read the new translations of the *Bible* into vernacular languages.

It must be noted that the second story also had its support from scholastic philosophy. The same Aquinas who contributed the most compelling statement of the theology of evil also wrote about possession and the powers of demons. Demons, he wrote, assail us by instigating us to sin, or are sent by God to punish us for our sins. Thus God *permits* devils to harass us, and the high ceremony of exorcism will be unavailing if the possession is a divine punishment. Furthermore, even those of our sins that result from our free choices (our moral mistakes) indirectly make us children of the devil, because in so sinning we imitate the first sin, that of Satan himself. Thus does theology support personification. Rational considerations show that what we know in our hearts and experience regularly is true: it is impossible for human beings not to sin.[6]

Where do they come from, the angels, the *disembodied spirits*? The ancient Hebrew tale tells that Jahweh made them late Friday afternoon, and did not

---

5 There was never a witch craze period in Jewish history. Torture was a common feature of trials of witches in the 16th and 17th centuries, but judicial torture to obtain confession is unknown in Jewish legal procedures. Furthermore, in cases involving the possibility of capital punishment, Jewish law disallowed the confession of guilt on the part of the accused. Such confession was of course often the only evidence in witch trials.

6 The infamous *Malleus Maleficarum* of Kramer and Sprenger is a good example of a failed attempt to harmonize the theological and the personification stories.

have time to give them bodies before dusk, the beginning of the Sabbath. Christians tell a different story, adding more substance to the Jewish myth than it can perhaps bear. We have Adam and Eve and the Fall, *and the temptation: enter Satan*. One angel turned devil is already on the scene. But one devil is obviously not going to be enough. We know that each of us is attended upon by two angels, one our guardian, the other, our tempter. Further, we are told that Adam separated from Eve, and lived for a period of at least 130 years, during part of which time they both consorted with unclean spirits (for example, Lilith, created out of the foulness and dregs of earth), thus creating the race of demons. It is not important that this tale of the origins of the devils is perhaps not consistent with the myth of the fall of the angels, of the expulsion of some of them on account of the prideful attempt to achieve dominance equal to that of God. That version of the story has it that the period between God's creation of the angels and the fall of some of them was very brief, perhaps in the interval involving the separation of light from darkness. No matter how the emphasis is placed, the important thing is that we have enough devils around to provide for all of the evil in the world. If attention to witchcraft is to be more than just the latest academic toy, it is essential that the extent of existential commitment to many demons be taken very seriously.

Well then, how many angels are there? In the 16th Century, a number between 7,405,000 and 7,409,000 was popular.[7] However, these numbers are clearly too small to account for the amount of evil in the universe. Better estimates are based on the assumption that there are 9 orders of angels, each with 6666 legions, and each legion with 6666 individuals = 399,920,004 angels.[8] According to *Revelation*, Satan drew one-third with him at the Fall = 133,306,668. Even if we make allowance for the size of the world population between, say, 1450-1700, this number of angels seems too small, especially when we remember that we are each attended upon by two of them. Perhaps the number can be seen to be satisfactory if we emphasize that unbaptized heathens are doomed to go to Hell anyway, and hence need no angelic protection and temptation. This exemption, however, would lead to the conclusion that Christianity ought not to become a missionary or evangelical religion, because increasing the number of Christians will eventually reduce and perhaps eliminate the available pool of angels. Especially dangerous is the threat of eliminating the 2-to-1 advantage the angels have over the devils.

Let us suppose that we can overcome these arithmetical difficulties and can arrive at an equation that satisfies the demands of the myth: angelic protection and satanic temptation. We can then provide a conceptual model for two kinds of

---

7 Johann Weyer, physician to the Duke of Cleves and critic of the witch hunts, estimated the number to be 7,409,127 commanded by 79 princes of darkness. A formula much in favor: multiply the Great Pythagorean Number (the Decad) by 6--1234321 x 6 = 7,405,926. 6 because the number of the Beast of the Apocalypse is 666 (*Revelation*).

8 It is of passing interest that this estimate is thought to be based on the composition of a Roman military legion Actually, a Roman infantry legion consisted of 10 cohorts each with 300 to 600 men. Demonological arithmetic is quite fascinating!

evident evil: the evil we ourselves perpetrate, and the evil others perpetrate. Evils I produce result from giving in to temptations of my satanic angel. My failure of will comes about – *we must take this quite literally!* – because I am bewitched, if you like, possessed. Exactly the same analysis applies to the evils perpetrated by others. In each case of evil doing, some supernatural intervention on the part of angels is essentially involved. The point to be stressed is that in each case of the initiation of evil, the person as perpetrator is not responsible for the evil act – *some demon is*. To be sure, the person as perpetrator has, as we say, *given in to* the devil. But where was the guardian angel in all of this? If my psychic life is a continuing struggle between two angels, both bent on having control of my soul, where do I fit in? Where is my freedom to choose?

Thus, I would suggest, the phenomenology of belief in witches reveals that an occult personal psychic drama is thought to be taking place in the hidden recesses of the lives of each of us. Evil results whenever devils get the upper hand. This is a story that can easily be assimilated when sounded from the pulpit; it personifies and renders concrete the inner struggle to be masters of our fate experienced by all of us. It tells us in vivid relief that we are, all things considered, merely instruments of beings we will never really comprehend. It also gives us grounds to understand how one can accuse someone of witchcraft. It has to do with scapegoating, with the deliberate attempt to transfer our own guilt, sorrow, and cosmic insignificance, to others. It has to do with the endless attempt to make others into things, those instruments of the devil who cause our plagues, kill our cattle, give us headaches in the night, render us impotent at just those times when we need an increase in the size of our families. It becomes easy to want to eliminate such instruments.

In order to solve the problem of evil at the level of actual human experience, Christianity created the witch. In so doing it introduced an existential social type: the human being debased as thing – *the other*. If I had more time I would welcome your reactions to the rest of what I have to say about the deep philosophical problems generated by witch hunting. We need to study the human purposive activity known as work to understand how its management generates norms of behavior operative throughout the provinces of civilization, in society, science, the economy, the arts, religion, and so on. The human correlate of a thing is a being who does not work, but is an instrument of the work of others. Think of witches, think of slaves, think of infantry soldiers. A major lesson to be learned from the witch craze of the Middle Ages and the Renaissance is that civilization cannot be maintained by means of dominance/dependence structures that turn dissidents into witches, workers into slaves, eaters of blueberry pie into infantry soldiers.

We have several theories of the origins of the medieval and Renaissance witch craze: Murray's thesis, the scapegoat theory, the feminist theory, the madness theory, and others. All of these theories, except for the scapegoat theory, take the phenomenon of the 16th- and 17th- century witch to be unique. If I am correct that the two stories of evil provided the conceptual background of the witch craze, then the witch can exist anywhere and anytime. It is all a matter of how we dress her:

she can ride a broomstick, work twenty hours a day in the cotton field, practice short-order drill to become an efficient fighting machine. The two stories gave the hunters the apparently justified will to hunt, and the hunted no place to go, for although they could not understand the theology of evil, they felt the horror of its personification. Their monstrous and hideous destiny has taught us that confession extracted by torture is not a rational alternative, and that silence makes one guilty by suspicion.

CHAPTER 2

DE PRAESTIGIIS DAEMONUM
Early Modern Witchcraft: Some Philosophical Reflections[1]

LECTURE 1

'WHO THEY BE THAT ARE CALLED WITCHES': The Ontology of Witches

The year is 1600. Much of importance is happening. Rubens is at work in Italy. Frescobaldi composes. The harp begins to be used in orchestras. Hundreds are entertained by Shakespeare's *Hamlet* and others of the Bard's plays. Construction of the Royal Palace in Naples begins. The recorder, wigs, and dress trains become fashionable. Henry IV of France marries Marie de Medici. The British East India Company is established. In Sweden, Catholics are persecuted. Half of Frankfurt's elementary schools are operated by women. In England, tobacco sells for its weight in silver shillings. Snuff and clay pipes become popular. The renewed interest in the consumption of tobacco saves the Colony of Virginia from financial ruin. French fur traders establish a post on the St. Lawrence. Three years later Champlain will occupy Newfoundland, Nova Scotia and New France. Science and technology continue their painful births. While Francis Bacon ponders the wonders of the methodology of the new science, Tycho Brahe and Johannes Kepler collaborate on matters astronomical in Prague. William Gilbert publishes his soon-to-be celebrated *De magnete*. Dutch opticians invent the telescope; the German

---

[1] [Previously unpublished]

Much of the work embodied in these lectures was made possible by the German Humboldt-Stiftung, in the form of a generous Research Prize. I thank the officers of this foundation for this tangible expression of support of my research.

Athanasius Kircher invents the magic lantern. Descartes, who will later – but perhaps too late? – introduce the new scientific rationalism, now in the care of his grandmother, is four years old. The event takes place that will serve as the emblem of the struggle between science and religion that continues to this day: Giordano Bruno is burnt at the stake for supporting the heliocentric hypothesis.

In late April of 1600 a family, the Pappenheimers, arrive by cart at the Falcon Tower, Munchen. There are five members of the family, Paulus and his wife Anna, their grown sons Michel and Gumpprecht, their 11 year old son Hänsel. In Worth, a small town between Regensburg and Straubing on the northern rim of the Danube valley, a thief by the name of Geindl had been hanged in the summer of 1599. Before he died, he stated that the 'tinker lads' (Michel and Gumprecht) had abetted him in the murder of 7 pregnant women. He said the sons travelled with their father and mother and younger brother stealing and killing wayfarers. The members of our family, then, are arrested on the basis of uncorroborated accusations. It is to be noted that incarceration in the Falcon Tower was reserved for particularly dangerous criminals, principally those thought guilty of *crimen laesae maiestatis divine*, high treason against the majesty of God, of which witchcraft was the main offense.

From the account in the old documents, we know a fair amount about the accused. Paulus had been a grave digger, beggar, 'emptier of privies'. Later, at fairs, he ran a small gambling table. In the summer he and his sons would make a modest living repairing broken windows. (The people of Nuremburg were to use the term 'Pappenheimer' for garbage collector right down to the 20th century.) Paulus, and later his sons as well, would clean the cesspits by lowering buckets on ropes and either throwing the contents onto fields or into running water. (The farms, middle-class households and castles tended by Paulus were relatively progressive so far as hygiene is concerned. By comparison, there was not a single privy in the Duke's Munich residence, and human filth accumulated in the hallways.)

The Pappenheimers were arrested under orders from Maximilian I, by the grace of god Count Palatine of the Rhine, Duke in Upper and Lower Bavaria, who was 27 years of age in 1600. They were tried, in inquisitorial fashion, by Johann Simon (or Sigismund) Wangereck, who was in his mid-30s, and a member of Bavarian Council of State. A few words about each of these persons of pivotal importance for the case are in order.

Maximilian had been subjected to the strict will of his father, and was eventually to become a degenerate psychological cripple in his dungeon of a palace. He was a hyperdevout Catholic, who thought of himself – not without justification – as free from Roman control, with superior privileges. Indeed, the major influences on him came from Spain. His courtiers wore Spanish styles of clothing; his views of jurisprudence (including the use of torture) came from the Spanish Inquisition. He had a morbid and growing fear of witchcraft, which he expressed often. It is quite likely that the innocent Pappenheimers were selected on his order to find good examples of witches to be held up to public view: scapegoats, the means of relieving oneself of one's own strong sense of guilt. It is to be hoped that Descartes, who at the age of 22 served in the Bavarian army in its fight against Bohemian Protestants, was unaware of Maximilian's deeply flawed character.

Johann Wangereck was a committed zealot, at times arrogant, at times unnecessarily accommodating. He was of lower Bavarian origin, uncouth, lazy, yet often overtaken by rage and misguided passion. He studied law at Ingolstadt (now best known for producing the Audi automobile). The university in Ingolstadt (founded 1459) had a thoroughly awful reputation: 'A good many parents would sooner permit their children to go off to war than to see them go to Ingolstadt'. Student life involved 'bestial guzzling and boozing', gambling, taverns demolished from sheer spite, general terrorizing of the populace. The students (of course all of them males) wore swords: brawls, stabbings, saber duels, were common, as were deaths. The faculty of law had only one moderately competent teacher. It was in such an environment of impoverished intellectual activity and constant violence that Wangereck received his doctorate in law.

The investigation of the alleged crimes of the Pappenheimers took the usual course of inquisitorial justice. Accusatorial law, involving tests of water, weight, and other measures of witchery, had for the most part been abandoned. The four adults were interrogated for various periods of time, Hänsel less frequently. As the manuals of procedure required, they were first told of the possibility of torture, were then shown the instruments of torture, and finally, were actually tortured. The law prohibited the use of instruments of torture on children; Hänsel was thus only subjected to being whipped! Under torture, the four adults admitted to having committed various crimes.

All four confessed to being seduced by devils and having made pacts. Anna confessed to having gone to the sabbath, a fact confirmed by little Hänsel. Paulus, Michel and Gumpprecht admitted to having stolen, murdered and put curses upon others. The inquisitors pressed hard on one strange point: *infant's hands*. Weakened and in intense pain because of thrashing with the rod, Hänsel confessed that his brothers had cut the hands off beggar children and had murdered pregnant women to obtain the hands of their unborn children. The significance of this: it was widely accepted that fingers could be used as 'thieve's candles', giving magical protection to the thieves. Infant's hands are particularly good for this purpose. The seat of immortality lies in the hands. So the infant hands possessed souls that were young and robust, innocent and free from sin, and unbaptized and hence of a magic nature. A powder of hands rendered as an ointment was another magical source of maleficient acts when smeared on the victim or an effigy of the victim.

Although mutual confirmation of similar points of guilt were elicited from the four adults, Bavarian law required that in addition some direct physical and perhaps eye-witness evidence of the alleged crimes be produced. On a number of occasions Wangereck sent deputies to look for corpses, check the torture-forced testimony of the Pappenheimers, and seek out any other useful bits of confirmation. Some enemies of the family did attest to their awful crimes, but no corpses could be found (of course!).

It was then decided, with the help of the *Malleus* and other sources, that since witchcraft is treason against the majesty of God, it is a crime the determination of which can proceed without the usual forms of confirmation. Accordingly, the torture-induced testimony of the five was taken as condemnatory. All that was needed was a constant watch of the prisoners – often the devil killed them in their sleep!

Note that at no time were the Pappenheimers represented by council. Elsewhere lawyers could be employed by those accused of witchcraft, and wealthy witches could sometimes buy their way to freedom by gifts of money or property. The Pappenheimers were dirt poor. However, defense lawyers, money and property did not matter all that much. So much happened in a trial exclusively at the discretion of the judge, who could simply ignore the defense plea, and who would in any case share the spoils with the Church and the State when the accused was executed. If the accused was wealthy, *she* paid the costs of her own trial and execution.

A guilty verdict was pronounced on the four adults. Some time between July 26 and July 29 (the date of execution) the barber arrived to cut off the hair of the accused. This was a ritual act, for hair was regarded as the seat of life, of the soul, of vital energy. Human hair is in the same category as devil's ointment and weather spells. Remember the image of the flying witch with free-flowing hair. Remember too that the Devil took hair from all parts of the witch's body as a necessity of the pact. It is not insignificant that the barber's fee was the equivalent of a year's wages of a Munich waitress.

On July 29 a procession a half-mile long brought the victims, now turned into mere objects, extras in this drama, to the place where the shrift was read. Guilty on all counts of witchcraft. The accused were given no opportunity to reply. Instead, in a dramatic act, the executive justice took the staff at both ends, and broke it. The execution could now begin.

Little Hänsel was forced to watch while the prisoners were stripped to the waist. Next, the executioner took red-hot pincers from a brazier and, one after another, ripped gaping wounds in the arms and torsos of the four. Finally, he cut off Anna's breasts and smeared them on her face and those of Michel and Gumpprecht.

As the prisoners approached the fire, they were required, according to ancient custom, to stop and to pray at a wooden cross adorned with flowers. They were offered large bottles of wine, which they drank eagerly. This act of compassion was a symbolic act, an institutionalized form of humanitarian sentiment. (Matthew 27:34; Mark 15:23: reports of Jesus's crucifixion; Luther: 'Give strong drink unto them that are to die'.) The truth: perhaps a drunken body can better withstand pain. Even a modest cynicism requires that one question motives here: the greatest gift of Christianity is perpetual guilt.

Paulus, Michel and Gumpprecht were then broken by the wheel. Paulus was impaled. In their final moments of intense agony the four were burned alive. Not even the customary mercy of a broken neck before burning was accorded them.

On the 26th of November 1600 little Hänsel and four who had given evidence against the Pappenheimers (not realizing that by so doing they were implicating themselves) were also burned alive.

So our story ends. Wangereck had done his duty, the Duke could enjoy a few hours of relief from the pains of hatred and unaccommodated guilt. Bavarians, at least for a short time, could take comfort in the illusion of living in a morally cleansed society. Is our story a typical narrative? Do all trials and executions of alleged witches in the early modern period conform to the model of the treatment of the Pappenheimers? Yes and no.

The burned Pappenheimers numbered four males and one female. It would be a serious mistake to generalize from this single case. In fact, the true proportions are exactly reversed: over 80% of those tried and executed were women. The early modern witch craze was truly the women's holocaust.

There are other prominent respects in which the case of the Pappenheimers is not typical. Whereas the German Dukes of the domains of the Holy Roman Empire possessed autonomous powers, and could usually exercise them without fear of serious interference, such was not true, for example, in France, where an aggressive, acquisitive and jealous king sought to impose a centralized elite culture on the popular culture of the peasants, singling out accused witches for special treatment. Nor was it true of the mountainous regions of Switzerland, Scotland, and south-western France, nor of the inland plains bordering Lake Constance, where Singen, then as in our century, stood as the very paradigm of social repression, intolerance, hatred and xenophobia. In these regions witches were born of folklore and tribal traditions. They may or may not have earned their reputation for promoting the commission of evil acts, but they were in fact healers and counselors. Their sin was an inability to read Latin coupled with too much concern for the well-being of others. Trevor-Roper and the 19th-century romantic historians created the myth that the classical witch hunts began in the mountains, domains populated by werewolves, vampires, elves and fairies, and all manner of evil-doing beasts. Arno Borst of Universität Konstanz has recently exploded that myth.

What is typical in the tale of the Pappenheimers is its substantive expression of a certain ontology of the early modern witch. Lying behind the alleged crimes of the Pappenheimers is the ever-present figure of Satan. The early modern witch is not a misunderstood goddess, nor like the witches of the Azande, nor the present-day witches of the Bocage in France, nor the blood-sucking witches of contemporary Mexico. These witches, however evil may be their doings, are not inspired to commit evil by the Devil. But the role of the Devil in human affairs is the key to understanding the classical witch. Such a witch commits *maleficium* under the direct influence of the Devil: she is in every important respect his slave. He may never be found when we look for him, but he is always there, always in wait for his next victim. The maleficence he produces through others is definitive of all evils except those brought about by individual moral fatigue.

The essentials of early modern witch-making are three. The witch makes a written pact with the Devil, signed in blood and consecrated by an act of sexual intercourse. The pact entitles the witch to secrets of sorcery, gives her the power of night flight, and establishes her as a permanent member of a coven attending the Witches' Sabbath. (Little Hänsel reports that all of this is true of his mother, Anna.) Much has been written – especially lately – about the sources of this taxonomy of the witch. And much of what has been written, I suggest, should not be taken too seriously. I have specifically in mind Carlo Ginzburg's attempt to find the myth of night flight in the ancient traditions of shamanism in the mountains of the far East. My own view of the matter is rather more simple. Angels can fly, and do so in the bright light of their eternity. Which only means that they, like electrons, can go from one place to another without ever having been anywhere in between. Devils are fallen angels, and thus can also fly, but of course in the dim darkness of *their* eternity, the counter-eternity to that of the angels. They pass on this quality of night flight to their witch victims. (Even Aquinas believed that though angels and devils are disembodied spirits, they can assume any material form they wish in order to further their aims in the world and to communicate with men and women. In their corporeal state they are capable of instantaneous flight from one place to another.) The only thing difficult to understand in all of this is how it is that non-literate peasants could have such a good conceptual grasp of nuclear physics! My hypothesis is that they were more attentive to the psychological moving forces in their dreams than we are. (Interestingly, the sociologist Monica Wilson has referred to the witch-beliefs of a group as 'the standardized nightmares of a group', in "Witch beliefs and social structure", *American Journal of Sociology*. 56 (1951), 307-13.)

So it seems to me to be the case that the stereotype of the witch comes directly from Biblical sources interpreted in accepted theology and from the introspectively well-understood terrors of dreams. The taxonomy is powerfully underwritten – and this must be stressed in all continuing efforts of understanding classical witchcraft – by traditions of male domination, and by monastic rejections of the natural functions of human bodies. Public rejections. Some of the medieval underground tunnels linking monasteries and nunneries still exist. (I will have more to say about the body of the witch in tomorrow's lecture.) For years, scholars have tried all sorts of tricks to prove that Catharism had non-Christian roots. Immanuel le Roy Ladurie has recently proven exactly the opposite to be true. For years, at least some scholars have tried to disavow the Christian context of witchcraft. The disavowal is unconvincing.

Our witch – we might refer to her as a *Christian* witch – is a creature whose identifying characteristics stem from a combination of self-exculpatory theology and an effort to communicate Church doctrine to the ignorant, illiterate masses by means of metaphor and personification. Augustine made the first move when he encouraged early Christians to try to understand that the gods and goddesses of the Romans are the devils of the *New Testament*. The social, political and psychological ramifications of this bit of theological trickery cannot be overestimated. Aquinas, of course, delivered the ultimate move in the intricate end-game of medieval theology by arguing that God not only creates Satan, but permits him to do his dastardly deeds through the agency of others. Why does a perfect, totally good, all-knowing and all-powerful being give such heavy, deadly, permissions? It is because good sometimes results from evil, and a perfectly good being could not want to create a universe in which such potentialities are never to be realized. It only remained for inquisitors like the authors of the *Malleus* to refine the procedures of discovery, trial and execution, and for Innocent XIII to issue the Bull of 1484 legalizing the trial and punishment of alleged witches, and the witch hunts could begin in earnest.

Even when evil is personified by identification with the devils, the mind-numbing theology remains opaque; after Kant destroys reason to make way for faith, and thus eliminates a merely empty theological word-play, the rationalizations of Augustine and Aquinas fail to satisfy. Evil is not an abstraction best understood by means of philosophical argument on the old style. Evil is something personal: it happens to *me*. The relationship of the witch to her victim must be understood as are other forms of human conflict involving an unequal distribution of power and influence. In addition, the relationship of the witch to those who hunt her, torture her, and execute her is a second form of conflict needing analysis. Tomorrow, I will try to delineate some features of these kinds of relationship.

LECTURE 2

THE WITCH AS OTHER, THE WITCH AS BODY:
The Phenomenology of Witchcraft

A witch is a socially and politically marginal non-person; she is the terrible other. So to speak of the witch is to engage in pleonasms: *to be* a witch is *to be* marginalized; *to be* a witch is *to be* the other. The witch, we are frequently told, is a cantankerous old hag, who at the very least is socially annoying, at the very most is destructive of life and property. A deeper investigation by Robert Muchembled[1] has revealed that her situation in the ancient village was in fact much more complex. Because she was thought to have special gifts (for the most part deployed in the interests of healing), she was feared and mistrusted at home. After all, her gifts could at any time be used in evil ways, to kill rather than to cure. However, the good villagers trusted witches resident in neighboring villages, because these witches did not know them personally. The relationship with a non-resident witch could then be more 'professional', and less threatening. The visitor could approach the foreign witch with a feeling of being 'in charge', a sense of being in control that could, however, quickly disappear if the witch so willed it. Consulting a witch, for whatever purpose and in whatever context, is always a very risky business.

Muchembled's work thus moves us to another level of understanding. And we can now ask: just what is it that characterizes the interplay between a witch and her victim, or her patient? So far as I know, the witch as healer prescribed drugs, suggested regimens of bodily exercise, burned incense, made potions and, in general, acted like a physician practicing natural medicine (German: *Heilpraxis*). It is the relationship between the witch and her victim, not that between a witch and her patient, that needs deeper analysis.

I want now to tell you a story. My tale is motivated by some ideas developed in Jean Paul Sartre's *Being and Nothingness*, especially the section on 'the glance' (*le regard*). I use the ideas as a springboard for further analysis; I would not want to be understood as subscribing to Sartre's form of phenomenology.

---

1 *Popular Culture and Elite Culture in France, 1400-1750*, translation by Lydia Cochrane (Baton- Rouge: Louisiana State Press, 1985).

I am sitting on a park bench on a patch of ground slightly elevated above its surroundings. I notice some geese on the nearby river. In a non-attentive mood, I just perceptually pass over trees, and grass, and clouds in the sky. I hear, but do not notice, familiar sounds. My sensory information, some of it consciously processed, some of it not, is all connected with objects, with what I might call *somethings*. *Things* is the word to emphasize. Apart from my consciousness of my own inner states, my perceptual experiences provide me with an *external world, a personal universe*. In an instant, my universe comes to be populated by what I recognize as not-me, a something with a body like mine, an *other*. As this body moves at a distance, I have no difficulty accommodating it to my universe of objects, the things of my world. Now the drama begins: the body turns about and its eyes glance at me, just a quick look, an *Augenblick*. My universe is suddenly transformed, I turn away, I do not look at the stranger, the other. For what the other has revealed to me is my essential aloneness in my personal universe. More potentially damaging is the fact that the other's glance fixes me as an object, a something, in the universe of another. This experienced debasement of my self in control is a form of consciousness I cannot tolerate. I cannot, except in moments of psychiatric torment, *want to be an object for some other*.

The glance of the other can be regarded as definitive of the concrete relationship between different individuals. This relationship is one of conflict, of power, of expenditure of energy determined to recapture one's personal universe. The other appears to me in the form of a *challenge*. The relationship, furthermore, presupposes human freedom. If I were locked into my personal universe, as the madman is, the glance of the other, although perhaps repulsive, would not constitute a threat; I would have nothing to lose, nothing to give up, and also nothing to gain. The unequal distribution of power would not matter to me at all. But a free human being (Sartre's being 'condemned to be free') must always fight for lost territory, must convert the glance of the other into a benign look by any one of a huge number of tactics, for example, sadistic reprisal, public humiliation, suicide, retreat into drug-induced reveries, infliction of direct bodily harm. We have all crossed to the other side of the street in order to avoid the glance of an oncoming other. This primitive tactic of conflict resolution (or avoidance), provides a phenomenological axiom for all forms of behaviour bent on maintaining an equilibrium of power.

The special case of the witch as other is to be understood as a theorem following from the axiom. The glance of the witch is especially terrifying. The evil eye always fixes me in a malevolent mode. To be bewitched is precisely to be rendered incapable of retaliation, to lose one's freedom, to come off second-best in the struggle for power. The evil eye, in short, makes me instantly into a thing with little hope of recovering the universe constituted by my inner consciousness. The devil does not want only part of me, he wants me whole and entire, body and soul. He wants to devour my freedom. So it is that the glance of the witch, the glance of the evil eye, comes to define the nature of the relationship between witch and victim. The evil eye fascinates, bewitches. Even presiding judges avoided being glanced at by those on trial for witchcraft, for fear of falling prey to the ways of Satan. What the Scots would come to call 'glamour' was well recognized in many cultures; the physician Johann Weyer lists some of the many remedies and preventive tactics recommended to defeat fascination: rue (Aristotle), bony date-pit, filed down by tooth, a branch of buckthorn, the herb Lady's-glove, the skin of a hyena's forehead, red coral, spitting into the fold of one's garment (Theocritus [*Idylls* 6.39]: 'I spat thrice into the bosom of my garment lest I be bewitched by the evil eye'; and Ovid [*Metamorphoses* 10.414]: 'Let each thus spit into the folds of his garment').

The psychic conflict involved between the various players involved in witchcraft has been studied by the contemporary ethnographer, Jean Favret-Saada, during long visits to the Bocage region of France. The expenditure of energy and power involved in the interplay between the witch, the bewitched, and the person who comes forward to unwitch the bewitched by bewitching the witch, is so electrifying and contagious that even a social scientist can become absorbed by it. No one understands better than Favret-Saada the general point I am trying to emphasize in the present analysis.

The peasant of the 16th and 17th centuries was no stranger to adversity. He had experienced plagues, failed crops, periods of virtual unemployment, periods of low prices for his products. Having been taught the myths about the devil and the features of the stereotype of the witch, how could such a non-literate person fail to accept that witches must be dealt with quickly and decisively? Our peasant understood perfectly that preservation of freedom entailed measures as awful as torture and execution, that at the level of concrete human experience conflict resolution can often only be achieved by actual elimination of the other. This is most precisely true of those relationships involving alleged witches and their victims. The painful evils inflicted in these relationships can only be made right by the continuing struggle to blind the evil eye. Theology, I strongly suggest, had nothing whatsoever to do with this form of human drama. In seeking to justify God's ways to mankind, Augustine and Aquinas and dozens of lesser thinkers may have thought they had argued evil away. But evil survives argument; it is resurrected in the daily struggle to achieve an equilibrium of power in human affairs.

In a society that has accepted the literal existence of devils and witches, the only way in which something like an equal distribution of power and control can be maintained is thus to destroy witches. The devil cannot be destroyed. Even Jesus, although he was able to resist the temptations of Satan, could not destroy him. Jesus was powerless to remove evil from the world. Instead, he offered himself as the savior, he offered eternal salvation in the next world, one in which there is no power and no conflict. It is thus the servants of the devil who must be destroyed, all of them, to the end of time. Many presiding magistrates, in moments of self-justification and guilt, actually thought they were doing the witch a favor by burning her body, because this would free her soul, and thus give her a kind of modified second chance.

The absurdity of this rationalization aside, there is a point here to be pondered. It is the point having to do with the witch as body. The other always comes into existence for me as body-of-the-other. However, Satan has not destroyed the body of the witch, he has corrupted her mind, disempowered her soul, disinfanchized her psychic potentialities. Why not then simply deal directly with the problems of the witch's bedeviled soul? The sacrament of exorcism was originally designed to do just that, remove the influence of devils from the body and soul of the afflicted. In the high period of the witch hunts, exorcisms were of course performed, but this enlightened form of removal of the influence of devils was largely replaced in practice by the inquisitorial process, involving torture and usually execution of alleged witches. This legal process introduces emphasis on the sins of the body, as we will see shortly. In doing so it provides a new model, different from that presupposed by the practice of exorcism.

In the case of exorcism, the emphasis is upon helping the victim, who is regarded as not responsible for devil-induced actions. The person bewitched is just a ventriloquist's dummy in this respect: a vessel filled with evil. The demons speak through her. God's aid (through Jesus) is invoked in the ceremony of exorcism, so she can be unbewitched. But the demons are not dispensed with, just removed from one body, and are apparently free to attempt to find other victims. Exorcism, then, is used to cleanse the soul by removing satanic irritants from the body. The procedures of exorcism rest upon a foundation of concern for the victim; the procedures are thoroughly Christian in the non-belligerent sense.

In the case of witchcraft, the witch is held responsible for her actions, even though she is bedeviled, and in fact is not responsible. The emphasis is not upon helping the victim, but in ridding the world of those who work for the devil. The witch, as part of the pact with Satan, has renounced God, so God cannot be called upon to help her. The only solution is to kill as many witches as possible, for all eternity. Presumably the death of the witch releases the victim from the evil spells, but this result does not necessarily follow. The devil is eternal = evil is eternal. Who knows for how long a witch's spell will last?

That such a major shift in models of treatment of witches could come about requires serious investigation of a kind historians are not likely to want to engage in. The historical narrative is clear. Originally, the legal position of the Church was based on the *canon episcopi*, a papal Bull recommending that witches should be avoided, in effect, turned into social pariahs, simply left alone. Serious cases (usually involving those who could afford the expense) were to be handled in the time-honored way by use of the sacrament of exorcism. This position was reversed by Innocent VIII's Bull of 1484, licensing the hunting, trial, and execution of witches. Even earlier, in 1398, the Faculty of Theology of the University of Paris had issued a Pronouncement listing all of the evils of the practice of witchcraft. The narrative only skims the surface. We need to face the facts of the important emphasis upon the body of the witch, indeed, upon any body, perceived as the revelatory structure of otherness.

Barbara Tuchman reports that in the 14th century one of the four most sought-after and eagerly-purchased papal indulgences was permission to follow the popular practice of having a body buried in two or more places.[2] (The other three, for those who are taking notes: legitimization of children (in a period of only one year, 484 out of 614 were dispensations to clergy), permitting nuns to keep two maids, purchase of priesthoods by illiterate men below the statutory age of 25.) To bury a body in two or more places: a strange custom, one contradicting the use of quartering a body to make a political or religious point, as in the case of Huldrecht Zwingli, the Swiss peasant-priest whose armed revolt against the Roman Church ended in his defeat and death in battle at Kappel, Switzerland. His body was cut into four pieces, and sent from village to village to remind the Protestants of their inevitable destiny. It was normal practice to confine quartering to those of royal birth. Zwingli's treatment was therefore exceptional, but understandable: he had converted six Swiss Cantons to Protestantism, five, the Forest Cantons, remained Catholic. The Swiss Confederation had suffered a serious division; the Roman Church felt compelled to act, as it always did under threat of loss of control.

There are good and sufficient reasons why the wealthy should want to buy this indulgence. There was a considerable *underground* business in robbing graves, both to sell the bodies for purposes of dissection to medical faculties in the universities, and to use human body parts in the many ways in which any animal body parts are used – for food, for making clothing, and for countless other household necessities. In addition to the grave robbers, wild animals had little trouble negotiating shallow graves, and could eat their fill. The trouble is that when the day of reckoning arrives, when resurrection occurs, one wants to have something to resurrect, some part hidden from and undetected by the grave robbers. By a curious inversion of spiritual values, the life eternal depends crucially on the once living all-too-mortal body. None of this is informed by and sensitive to the Pauline doctrine that on the day of eternal reckoning the body that will 'arise' is a *pneumatic* or spiritual body, not the once lived-in material one. Paul's reliance on the Greek notion of the soul as air was just too exotic for popular Christian consumption, and in any case was one whose contorted internal logic largely escaped his own otherwise keen intellectual perception.

The customary caution of multiple burial sites was not accorded the witch whose body was burned. The witch could be excommunicated and anathematized, punishments that in effect sealed forever her possession by devils. She could be exiled, cut off from family and friends, if any survived. But to have her body destroyed utterly, although it may free the spirit, frees a spirit that will never in all eternity have a habitat, a place of *corporeal* rest. How did the body of the witch acquire this very special status?

We need once more to study some theology, but this time some heretical theology. I hope you will share with me the joys of visceral titillation brought on by the change in perspective. Jesus, so the standard story goes, died for us on the cross; his body, what many thought to be the body of god, died for us on the cross. That very same body reappeared after death, and ascended into heaven, taking its

---

2 *A Distant Mirror: the Calamitous Fourteenth Century* (New York: Knopf, 1978), p. 27.

rightful place in an untroubled eternity. The emphasis upon the material body of Jesus is so crucial and fundamental as to reveal the very essence of Christian commitment. During the celebration of the sacrament of the Eucharist, the power of transubstantiation is so great that it converts the wine into actual blood and the bread into actual flesh. As a matter of faith the idea of transubstantiation (and of its rival, consubstantiation) is not problematic. As an idea to be understood, it has evaded the best attempts at non-trivial explanation. It is an idea that philosophically divided Luther and Zwingli, thus preventing what could have been the most important creation of Protestant solidarity. It is an idea whose rejection has led to some intriguing theologies eventually declared heretical by the Roman Church.

Docetism is the doctrine that Jesus only appeared to have a human body, but that in fact he did not, and could not, have had an actual human body. Why? Because the human body is conceived in sin. Bound eternally by the conditions of original sin, it is impossible for a human being not to sin. Even the lives of the saints are not free of sin. The Son of God, or God himself, cannot be morally limited by such a body, one festering in moral failures of all kinds, an endless pit of corruption and deceit. The Gnostic Christians, who flourished in the early centuries after the death of Jesus, were docetists. So were many other early Christians. Their doctrines were all declared heretical by the Church. The Cathars, who lived in the Languedoc and were at full strength in the 12th and 13th centuries, were perhaps the most extreme docetists of all time. Perhaps that is why they were totally destroyed by crusades authorized by the Popes in Rome. So much did they detest the human body, as the seat of all sin and all forms of sickness and corruption, that they hoped for the eventual suicide of the human race. They extended this attitude about the human body to all corporeal creatures, refused to eat meat, and lived under an interdiction that men not touch women, substituting a form of spiritual marriage for carnal marriage. They believed in the transmigration of spirits, hoping that their souls could be free of human contamination if they were released on death and were able to fasten on to a rapidly moving creature, like a horse running by. To be reincarnated as the spirit of a higher life form free of the weight of guilt was their abiding goal.

Jesus, they thought, was a different kind of being. He was not subject to the limitations of the human or any other kind of material body. He brought a simple message of love and concern to human beings, but, unfortunately, human beings are, as embodied, incapable of living the kind of life he recommended. For our purposes, it is the Cathar stress upon the moral and spiritual unworthiness of the body that is of cardinal interest. In large part, this idea is entailed by their theological dualism. They believed, as did the earlier Gnostics and the young Augustine, that the universe has two sources, one good, one evil. The good part of the universe is spiritual, the evil part is material, corporeal. God and Satan are co-authors of the universe. Again, it follows logically from this dualism that Jesus could not have had a human body, for the incarnation would immediately Satanize him. Unfortunately, the Cathars shared the prevailing practices of male domination. Women were thought of as not much more than cows. Some male Cathars even believed that women's souls cannot be received into Paradise until they are reincarnated as men. The net effect of this chauvinism, intended or not, was to place women in that part of the universe created by Satan. Hence, although their accepted religious doctrine was one that recommended something akin to celibacy, male Cathars quite frequently violated sexual taboos, having intercourse with female relatives, and engaging in seasonal orgies. The justification of this deviant behaviour lay in the inherent fatalism of their theology, and in their rejection of free will: what will be, will be.

The moral uncleanliness of women was not only stressed by heretics, it became a central theme of the developing official Christianity. The celibacy of monks and priests was supposed to be based upon a certain low moral status of women: after all, it was Eve who was tempted by Satan in the Garden. The body of the woman comes to be viewed as a terrible thing, as a thing expressing the very antithesis of the worthy spiritual life: as the body of the other. There we have it: both orthodox and heretical Christianity marginalize women directly and with what they regard as strong philosophical and Biblical justification. And wherever a woman is, Satan is not far to seek. Why? Because women are by nature sexual, rather than spiritual, beings. Kramer and Sprenger, the authors of the *Malleus*, say it quite directly and without equivocation: women are sexually insatiable. Which means they will stop at nothing to tempt men into sexual activity, and which even more importantly means they will be easy marks for the devils, hungry as they are for as much sex as they can get, even with beings possessed of ice-cold penises. Against this background of structures of male-domination (there are many more than I have had time to mention), the body of the woman becomes, almost by definition, the utensil of the devil. The body of the witch is the body of all women, women made into eternal strangers by being essentially endowed with male-manufactured satanic features.

Exodus 22:18 – 'You shall not allow a witch to live' – became the battle cry. That the devil and personal sin are originators of evil is a given of the sacrament of baptism: '... *he* shall not be ashamed to confess the faith of Christ crucified, and manfully to fight under his banner against sin, the world, and the devil...'. 'Christ's faithful soldier' is masculine. In the Middle Ages the fundamental right of the free citizen was the right to bear arms, a tradition with a long future history. Jews, and women, did not possess that right. The war a man must fight is a war against his own inclinations to sin, against the temptations of the devil, and against the world (recall, the Devil is the 'prince of *this* world', so the world we live in as mortals is an ever-enlarging group of enchanted females.)

The great French historian Jules Michelet blames it all on the doctrine of original sin.

> One is filled with amazement to see all these widely different epochs, all these men of varying cultivation, unable to make one step in advance. But the explanation is simple; they were one and all arrested, let us rather say, blinded, hopelessly intoxicated and made cruel savages of, by the poison of their first principle, the doctrine of Original Sin. This is the fundamental dogma of

universal injustice...[3]

What is this so-powerful dogma, and what are the implications of accepting it? The original sin is the sin of pride, it is Eve's and Adam's prideful expectation that they can learn to know as much as Jahveh knows. They eat the forbidden fruit, after, let's be clear about this, Eve has been seduced by the demon snake. The punishment is swift and unusual: Eve and her female progeny will suffer painful childbirth; she and all subsequent women will be forced to accept their husbands as their masters. The punishments are not gender-symmetrical: all that is required of males is they work for a living. Jahveh also reveals to Adam and Eve a remarkable fact: although hitherto unaware of it, you are both naked. Shame and the rejection of the natural animal condition of humans are introduced conjointly. Thus welded together, they will in time provide Christians with the foundation of the omnipresent feeling of guilt. To be guilty about one's own nakedness before the body of the other. To want to hide one's animality. To come to regard sex as forbidden, requiring it to be *glamorous*. To seek a target to release the same – the body of the other, the body of the fascinating woman, the being who with one darting enchanting glance can steal your universe. The guilt becomes too much to bear. For too many, it becomes the origin of otherwise inexplicable violence against women.

The phenomenology of the witch as body reveals rearranged relationships. It is no longer a matter of the dynamics of the relationship between the witch and her victim. We now have to come to understand the dynamics of the relationship between the witch as body and those who would hunt her, interrogate and torture her, punish, and in many cases execute her. Because the witch as body lives in a morally deficient state, she is potentially the other to all males. Again it follows that the only way to maintain a favorable balance of power is to limit the sphere of activity of all women. The body of a woman is a body that must die, if not biologically, then surely socially and politically. The bigoted fanatics who burned the bodies of witches burned the bodies of women. They failed to draw any distinction in this regard.

---

3 *Satanism and Witchcraft: A Study in Medieval Superstition*, translation by A. R. Allinson (New York: Citadel Press, 1962), p. xii.

LECTURE 3

CAUSES AND SIGNS: the Epistemology of Witchcraft

In the Preface to his *De Magoram Daemonomania* (Basil 1581; Preface: Laon, December 20, 1579; 10 editions by 1604) Jean Bodin states that often one is justified in believing in the existence of something, even if one does not know the cause of its existence. The remark must have been acceded to by his contemporaries as referring to a matter that is beyond dispute. They would have stated, if pressed: we believe in the existence of God, who indeed has no cause, or is the cause of himself, even though in the first case we have no explanation of his existence, or, in the second, are left with an insoluble enigma.

As a young graduate student, endeavoring with difficulty to understand existence without causation, I attended a meeting of the Eastern Division of the American Philosophical Association to which J. B. Rhine of Duke University, then the *Magus* of studies in paranormal psychology, had been invited. He told us that in what he took to be carefully controlled experimental situations, there were those who were able to guess which card will be called up next from the deck, and this with a relative frequency exceeding chance. What explains this difference between such persons and ordinary mortals doomed to apparently permanent stochastic limitations? The privileged guessers have extrasensory perception. But why should they have ESP and not I? Rhine was caught in a messy philosophical trap. The best he could do was to repeat the message of the experimental results. ESP exists in some persons; that fact is clear and established. Even if we do not stress the obvious point that Rhine's argument is circular – the best guessers have ESP, and they have ESP because they are the best guessers – something is lacking here by way of causal explanation. What we have instead of causal explanation is an example of Bodin's belief in existence without known cause.

Unlike Rhine, who positively adored the presumed existence of ESP, Bodin was deeply troubled by something that exists without a known cause. The something is the result of magic, sorcery, especially that *maleficium* practiced by witches acting under the influence of demons. Bodin's view regarding existence without causation implies that the results of sorcerers' acts are unique, individual events, each one of which has never occurred before. Are there then not *types* of sorcerers' acts? Yes, there are, and Bodin goes to great lengths to provide a classification. But the taxonomic types are best thought of as grouped only under words, they are nominal types, not real classes, not natural kinds. A cunning move by the genius of the way with words. For, words aside, each act of magic brings about a novelty. The causation involved, if we may so speak, is rationally impenetrable supernatural causation. Which means: no time to transpire, no space to traverse, only the coming into existence of something novel: the death of *this* cow, *these* blotches on the face of *this* child, the infertility of *this* woman, *these* hailstones hitting *this* field. Each of them an event not hitherto in existence. (Miracles are thought of as violations of natural laws. A miracle is a causal anomaly, a novelty, and is generally accepted as something good. Hence we might take acts of sorcery, by inversion, to bring about *evil miracles*.)

I will not dwell on the problem of ontology here revealed. But I should mention that strictly speaking every event, magical or not, is novel, unique. It is when events get trapped into types that novelty and individuality begin to disappear. When this happens, we are urgently in need of poets, those liars by profession whose magic creates unclassifiable happenings.

If, as the analysis of Bodin suggests, we can never catch the witch in the act of devil-induced evil, if we can never trace an event to a given witch as the cause of that event, how are we to deal with alleged witches? How are we to discover that this woman is in fact a witch, and under what force of law are we to bring her to trial, in many cases exacting that most awful of punishments, death by burning? If analysis of causes, the search for what we now call physical evidence, is unavailing, what's left? Does the evidence against a witch always have to be circumstantial?

By the end of the fifteenth century jurists both ecclesiastical and civil had found ways to circumvent these problems. Many scholars have suggested that the key factor is the replacement of accusatorial law by inquisitorial law, whose procedures made torture legal. Although it is part of the truth, taken in isolation, this suggestion will not do. We still have to know how the judges selected this person for torture, rather than another. A vast body of penal law, much of it crafted in the images provided by authors of manuals like the *Malleus Maleficarum*, was developed to aid in persecution, torture, and execution of those presumed to be witches.

What has not, to my knowledge, been adequately emphasized, is that the penal law of European jurisdictions of the sixteenth and seventeenth centuries was legitimated by being underwritten by an obvious, silently assumed, but never explicitly acknowledged, epistemology. This epistemology replaced *causes* by *signs*, and in so doing substituted the attempt to show that an accused witch fitted a preconceived pattern for the search for causal links that would prove or disprove her guilt. It is not that the jurists working at the high point of the European witchcraze did not know about causes. They had studied Aristotle. It is rather that, following Bodin, they for the most part accepted that although supernatural causation fails to *explain* the existence of witches, what needs to be accomplished is a careful study of those signs that characterize someone as a witch. That witches exist was taken as a given; what a person (usually a woman) looks like when compared with the images of the current stereotype of the witch is the only question needed for purposes of making law.

Permit me to review some obvious features of the contrasting logic of causes and signs. Causes must be discovered; their whereabouts and whenabouts need to be ascertained. The best instruments we possess for determining causes are the methods of the physical sciences. In the most reliable cases the discovery of genuine causation yields explanatory laws of nature. As we have been reminded recently by media coverage of the O. J. Simpson trial, even the presumption of legal guilt must wear the costume of physico-chemical evidence. The lawful habits of DNA structures become pertinent in the matter of determining guilt or innocence. In the most important issues that face us – matters of life and death – we have come to trust the causal laws discovered by the physical sciences.

Signs are altogether different beings. We know where and when a given sign is. Discovery does not determine signs, presence at a place and time does. Understanding causation is an empirical task; understanding signs is hermeneutic. Explanation by reference to causes ideally involves appeal to natural regularities we take to be lawlike. Explanation of signs involves interpretation: to signal is to invite semantic assent, agreement as to what the sign means, if a symbol system as a collection of signs is to have fixed meaning. Semantic stability is not determined by appeal to laws available in a domain that is in principle public. Instead, semantic stability is assured by appeal to authority. In principle, anyone with the wit, intelligence, training and occasion can repeat the experiments providing evidence for a natural law. Such publicity of access is the hallmark of the empirical investigation of nature. Understanding signs is different. In many cases we consult rule books: dictionaries, encyclopedias, rules of the road, maps. Driven to extremes, we might also consult astrological charts, those who can 'read' bumps on the head, aroma therapists, wizards and divines. Once again: for the language of signs to work, the semantics giving it meaning must be privileged, authoritative.

Appeals to authority ought not to be ignored. They are essential ingredients in the epistemology of reading signs. In matters pertaining to witches, they can endanger life and limb. It is time to see how all of these analytic considerations apply to the case in hand: legal trials of persons accused of *maleficium*, of practicing witchcraft.

Medieval accusatorial law was fairly simple, if also fairly corrupt. One brought a charge against someone else by accusing that person of a certain presumed wrong-doing. When dealing with accusations of *maleficium*, magistrates always had the body of the alleged witch examined to see if the devil's mark was anywhere visible, a procedure that often involved shaving off all visible hair from the body of the accused. The accused was also subjected to one or more 'tests', negative outcomes of which indicated guilt. There was the test of water, of hot coals, of the heavy Bible, of being weighed. The justification of the evidential strength of the tests rested on the belief that it was God who was making the decision on the outcome. (Yet how easy it was to buy off the magistrate and have the accuser found guilty of error in the accusation, and thus held liable to pay the penalty of the talion!) Procedures of inquisition employed the tests of accusatorial law. Although the magistrates now took upon themselves the weight of decision, they could lighten that weight considerably by bearing in mind that, after all, the witches they condemned to death did 'freely' confess (after body- and mind-destructive torture).

During the reign of Emperor Karl V, in 1532, in the Reichstag zu Regensburg, a new penal code was introduced, one whose procedures for assessing evidence were borrowed from the Inquisition. It was called *Constitutio criminalis carolina*, and was the law in the Dukedoms of the Holy Roman Empire. An important part of this code deals with the question of when the evidence against an accused (especially a witch) is sufficient to warrant torture (*Indicia suffientia debent praecedere torturam*).

As noted, the prosecution of the alleged witch takes the inquisitorial form of moving toward a confession, normally extracted by torture. It was well known that few accused could withstand the physical and psychological damage inflicted by torture. Sometimes the torture killed the subject (an event always blamed upon the intervention of Satan), only rarely did the subject actually survive the torture without confessing to whatever crimes the magistrate cared to list. So the question of evidence sufficient to legitimate the initiation of torture is of the greatest importance. In the *Carolina*, it is a question framed in the terminology provided by the epistemology of signs.

The question has to do with what are called *indicia*. The *Carolina* classes as *indicia* the following: '*praesumptio, conjectura, signum et suspicio*'. These hypothetical states all provide grounds for inferences of guilt. For example, the sense of *signum* intended refers either to a physically embodied sign (for example, the Devil's mark) or a fact disclosed by the testimony of witnesses, two of whom are normally required. Some *indicia* are so evidentially damning that only one suffices to initiate some aspect of legal prosecution. For example, the daughter of a woman charged as a witch was summarily tried and sentenced as a witch.

Normally, a trial involves three parts. There is, first, a '*generalis inquisitio*', which precedes the '*specialis inquisitio*', a prosecution that can involve both torture and sentencing. The general inquisition attempts to establish the *corpus delicti*, which in most cases is so difficult that the *Carolina* empowers the presiding judge to use his own judgment in determining whether or not a trial should proceed. But even in the absence of physical evidence, he needs *indicia* sufficient to begin the special inquisition. Such *indicia* are given in the form of complaints against the accused, evidence of ill fame (a bad reputation in the community), or simple denunciation.

If the magistrate is satisfied that the *indicia* disclosed by the witnesses are sufficient, he will proceed to the special inquisition. For torture or for condemnation, it was normally required, in criminal cases, that there be absolute proof of the *corpus delicti*. As we have seen in the case of the Pappenheimers, such proof is difficult to obtain, and, if the accused is suspected of witchcraft, which as treason against the majesty of god is the highest of all crimes, establishment of the *corpus delicti* can be waived. Instead, the judge can use the signs provided by the accusatorial tests, or other signs introduced into evidence by witnesses. In this respect, possession by the accused of the *stigma diabolicum* or devil's mark is normally taken as infallible evidence of witchcraft. The mark, about the size of a pea, will not bleed when pierced by a needle, and does not feel pain. Since such a mark can only be obtained in the celebration of the pact with the devil during the rites of the Sabbath, it marks the accused as a witch with certainty.

When the judge thinks the *indicia* presented by witnesses, by tests, and by examination of the body of the accused, are sufficient, he will order the commencement of torture. Under torture, the alleged witch will be asked questions from a prescribed list. The questions will be repeated many times during the special inquisition. Consider the following list of questions legislated by the Catholic administration of Baden-Baden in 1588. The trials in which such lists were employed occurred in civil courts, although by 1588 the Vatican had given a blanket permission for the ecclesiastical authorities in the Holy Roman Empire to intervene at any stage of the proceedings. The list is typical of both Catholic and Protestant jurisdictions of the time.

The heavy emphasis upon matters sexual in the list of questions indicates again an explicit employment of a certain mythology concerning the body of the alleged witch. We must also note the clever way in which the questions are phrased. They are leading questions: a 'yes' answer to any one pretty much implies a positive commitment to all the others. One would expect this kind of guile; after all, it is now assumed that the accused is guilty of witchcraft. All that is needed in order to proceed to sentencing is a confession.

The list also displays all of the key ingredients of the stereotype of the Christian witch. During the interrogation, the answers of the accused will never be checked. When, rendered physically and psychologically dysfunctional by the torture, the accused begins to answer all questions in the affirmative, she is not supplying evidence for which the court needs collaboration, she is filling in the details of a well-understood stereotype whose semantics is fixed by folklore theologically clothed. Her 'yes' is an *indicium*, a *signum*, of a certain meaning to be inferred, an interpretation to be revealed.

I believe it is fruitful to view the Baden-Baden list as a classification scheme, a taxonomy of traits that are possessed by a witch. The list, if you like, *defines* what it is to be witch. Each question provides an image of something witchlike; taken together, the images are synthesized by the overarching image of the witch. However, although the list provides pattern and order, synthesis and definition, *it explains nothing at all*. The concept of the witch is not an explanatory concept. The list provides no confirmed or probable theory of witchcraft. If the accused fills in the pattern with 'yes' answers, this determines that she is a witch (which was the assumption accepted all along). But nothing that happens in the trial, especially the raising and answering of the questions on the list, proves that the witch *caused* anything at all.

It is not unusual for systems of classification to involve concepts that are not explanatory. Psychiatric nosologies have been atheoretical since their introduction in the 18th century. The *Diagnostic and Statistical Manual of the American Psychiatric Association (DSM)* even thinks it a merit of its system of classification that it is atheoretical, grouping symptoms in order to define mental disorders in much the way in which the Baden-Baden list groups images in order to define the witch. The *Manual* argues that, because its nosology is atheoretical, it can be of use to psychiatrists holding differing theories of the etiology of mental disorders. In much the same way one might argue that the atheoretical, non-etiological taxonomy provided by the Baden-Baden list was useful because it provided a system that could be used by those accepting different religious beliefs (both Catholics and Protestants burned witches), and also could be used in different social and political contexts.

It is particularly interesting to note that whereas the taxonomy of witches is based upon the authority of religious and civil institutions, variously expressed in manuals like the *Carolina* and the *Malleus*, the authority for the classification of mental disorders is statistical, based on average numbers of symptoms found empirically to be associated with any given disorder. However, the process of fitting an individual's behavior into a statistical pattern is epistemologically just like fitting an individual's behavior into a set of images synthesizing the concept of a witch. In both cases, no etiology is involved, no causal regularities or natural laws are appealed to, as, for example, in the case of the taxonomy involved in the periodic table of the elements, where assignment of an atomic number depends upon the recurrence of physical and chemical properties in accordance with the periodic law. Moreover, as is generally true of classification systems based on laws, the periodic law correlates many properties of the elements, including physical properties (for example, melting points, densities, crystal structures, hardness, thermal conductivity) and chemical properties (for example, reactivity, valence, polarity, solubility).

The periodic table of the elements classifies elements as natural kinds, types of objects subject to natural law. In the absence of such lawful support, the mental disorders classified in *DSM* and the concept of the witch qualify as a very different kind of knowledge. 'Witch' is in a certain sense just a name, as I suggested in my earlier analysis of some views of Bodin. Similarly, 'histrionic personality disorder' is also in a certain sense only a name. In the 18th century Kant suggested that physicians can help their patients by giving their diseases names. He proposed what he called an *onomastic* of diseases, a list of names for clusters of differing symptoms. The Baden-Baden list of questions and *DSM* provide onomastics, nominal designations, in the absence of lawlike information that could yield more substantial kinds of knowledge.

These matters of theory of knowledge and logic are ingredients in philosophers' games. Yet they reveal something that is not at all a matter of play. For the deadly serious concern we should all share about the epistemology of signs and symptoms is the question of the moral, social and political status of authority. We live at a time when, unfortunately, I think, the authority of science is being undermined by a host of competitive forms of belief. Some have rejected that authority outright. What we need to consider seriously is just what credentials the competing interpretations can appeal to, and what it is concretely that they have to offer.

There were periods when, on the authority of this or that prince, witchcraft was treated as heresy, and witches were burned because they rejected the one true religion. But witchcraft is not to be confused with heresy, if by 'heresy' we mean what was meant by this term in the Middle Ages:

> an opinion chosen by human perception, founded on the scriptures, contrary to the teaching of the church, publicly avowed and obstinately defended.

The witch is an apostate, not a heretic. She has rejected all of the doctrines of the Church, and forms her 'opinions' in accordance with the dark teachings of Satan. She has rejected both scripture and god. The heretic defies authority, as did the Cathars, by corrupting one or more of the teachings of the Church. The witch defies authority by becoming the willing tool of the counter-authority, Satan.

This much granted, there are still similarities linking witches and heretics, indeed, linking any who, for whatever reasons, reject 'official' authority in ways that appear to threaten the hegemony of those in power. 'Witch-hunting' is a phenomenon with a social and political structure, a set of conditions applying to heretic-hunting and to harassment of those who accept 'politically incorrect' views. The structure I have in mind involves relations of social and political authority and acquiescence, dominance and dependence. There is what seems to be a never ending urge for some humans to want to establish what I will call 'cognitive monopolies', dominant and authoritative social and political structures that completely determine what is properly to be known and believed. The victims of these monopolies are legion: Jews and heretics, those who accept new, strange and seemingly counter-intuitive ideas, communists in McCarthy's U.S.A., witches, old-fashioned university professors.

When a cognitive monopoly is threatened, preparations for retaliation take the following form. First, there must be an *authorized and organized power group*, which consists of the defenders of the cognitive monopoly, including those who operate the major political institutions, practitioners in the legal system, secret agents. Second, *officially approved doctrines* must be in place. Third, a *target group* thought to be unusually prone to act out of accord with the approved doctrines is required, a condition easily fulfilled by precisely that group whose views are assumed to threaten the cognitive monopoly. Finally, fourth, *unequal checks and balances* on the behavior of members of the power group and the target group will have to be characteristic of the relationship between those groups.

In the case of persecution of Jews and heretics in the Middle Ages, both prominent target groups, the official power group was of course the Church in Rome, supplemented in time by civil authorities, including those applying the legal codes. The official doctrines stemmed from the Bible, theologically and practically interpreted. That there existed unequal checks and balances on the behavior of those in the power group and those in the target groups goes almost without saying. The Jew, the Eternal Stranger, stood outside the social and political system altogether, and the same is largely true of the various groups marked as heretical.

The structure of resistance to dissidence applies in pretty much the same way to the persecution of witches. The one item requiring emphasis has to do with the matter of official doctrines. It is true, as we have seen, that persecution of witches was authorized by Papal Bulls, was justified by appeal to theologically interpreted passages in Scripture, and was carried out in accordance with procedures detailed in legal manuals. But more must be said. The witch, as apostate, rejects the ultimate authority, the word and existence of god. It must have been comforting to those who persecuted witches to have absolute assurance that they were doing god's work. Besides, the work became lucrative as time went by, as the hunters selected more wealthy victims, and as a whole industry of witch-hunting, witch-trying and witch-burning came into being.

I leave it to you, faithful listeners, to supply your own examples of the cognitive monopoly under siege. My own best example, because it was a period in history I lived through and whose pains I felt directly, is one that involved a *merely imagined, politically motivated* belief that the monopoly was under attack. I refer to the witch-hunt for those assumed to be communists mounted by Senator McCarthy in the 1950s in the United States. During that period the U.S. Congress lawfully changed a line in the old Salute to the Flag: 'one nation indivisible' became 'one nation, under God, indivisible'. Witches, witches, everywhere. But the telling of this story is best reserved for another occasion.

Philosophers should not preach, yet I am afraid I have in a way been doing just that. I suppose, because of all the sex involved, one can justify a prurient interest in witchcraft. Or because of all of the dusty old documents, a historical interest in witchcraft. Or because of the continuing struggle for women's rights, a feminist interest in witchcraft. Or because there are still witches at work in present-day societies, a sociological or anthropological interest in witchcraft. Or because the world of witches is a world of signs, a philosophical interest in witchcraft.

I confess to having these interests, and more. But above all, my interest in the phenomenon of witchcraft has to do with maintaining the values of human freedom in a world of unequally distributed power, with providing the cognitive resources necessary for the making of rational choices. If we fail to study and to philosophically ruminate on the dark side of human history, we lose perspective when confronted with present evils. For we, not the witches, are the authors of evil. By means of a conceptual transposition hitherto known only to brutalized witches and other tormented dissidents, we are today losing our hold on the idea of the *victim*. We live in a no-fault universe. How convenient it is for everyone to be innocent. How forgetful of the dark side of human history.

PART II

PHILOSOPHY AND SCIENCE FROM LEIBNIZ TO KANT

CHAPTER 3

LEIBNIZ' MONADS: A HERITAGE OF GNOSTICISM AND A SOURCE OF RATIONAL SCIENCE[1]

A typical view of Leibniz' extraordinary metaphysical and methodological views is to regard them as having appeared by certain "applications" of his mathematical and physical discoveries. If we believe Couturat and Russell the monadology is largely only the expression of certain logically formal relationships: it is the logic of Leibniz that is basic to his mature metaphysics.[2] Even more typical is the view that Leibniz' work in mathematics, especially his work on the calculus, is the source of the key ideas in his metaphysics. J. M. Child, the translator of the early mathematical manuscripts, says it straight out: "... the main ideas of [Leibniz'] philosophy are to be attributed to his mathematical work, and not *vice versa*."[3] In a philosophically much more interesting way Paul Schrecker argues that what he calls Leibniz' "infinitesimal method," a method obviously derived from his mathematical work, pervades all of the important basic contributions of both his metaphysics and methodology. The infinitesimal method involves a rationalist postulate expressing the view that ordinary modes of thought (and ordinary modes of language) be replaced by rigorous formal systems of expression and calculation. his work in logic, especially symbolic logic, and his mathematical investigations

---

1 [Previously published in *Canadian Journal of Philosophy* 10 (1980) 47-62.]
An earlier form of this essay was read at a workshop in history and philosophy of science held at Benmiller, Ontario, in May 1977. The workshop was supported by the Department of Philosophy, the University of Western Ontario, the Canada Council, and the Division of Logic, Methodology and Philosophy of Science of the International Union of History and Philosophy of Science. Further work on the material was encouraged by Jürgen Mittelstrass, Thomas Lennon, and John King-Farlow. I owe much to former teachers: Paul Schrecker, Francis Clarke and Milton Williams, immortal monads who have found new places in Being.

2 Louis Couturat, *La Logique de Leibniz d'après documents inédits* (Paris, 1901); Bertrand Russell, *A Critical Exposition of the Philosophy of Leibniz*, 2nd ed. (London, 1949).

3 J. M. Child, trans., *The Early Mathematical Manuscripts of Leibniz* (Chicago and London, 1920), p. iii.

led him to seek to extend this postulate to the entire area of scientific endeavor. For science to progress, this postulate must be adopted in order to conquer the unfortunate excesses of intuition and imagination. the second feature of the "method" is the celebrated principle of continuity, a principle usually thought to be a direct consequence of Leibniz' abiding concern with the question about how many numbers there are between some definite nonzero number and zero.[4] [I have deliberately put this in a fairly abstract way. The question of *quantities* or *magnitudes* that are infinitely small remains, I think, unanswered.]

Schrecker's persuasive analysis leads him to the conclusion that in the *Monadology* "the entire metaphysical system presented there represents the most profound and comprehensive application of the infinitesimal method. If it is not interpreted in this sense, it can hardly claim to be more than a sort of platonic myth."[5] [In what follows, he allows that the methodological trappings are nevertheless present.]

More recent attempts to deal with the elusive sources of Leibniz' contributions to the seamier side of science stress what are referred to as the "analogies" linking ideas in his positive science and his metaphysics. But even given this generous new reading, it appears typical for the analogies to be viewed always as moving *from* the scientific concepts to the monads. In their entry in the *Dictionary of Scientific Biography*, Mittelstrass and Aiton, after brief mentions of the roles played by the idea of force and the idea of organic growth in the thought of Leibniz, conclude: "It thus appears that physical analogies very probably provided the initial inspiration for the formation of Leibniz' metaphysical concepts."[6] But of course it is Gerd Buchdahl who makes most of the powerful analytic tool based on thinking of the relationship between metaphysical ideas and scientific concepts as analogical. Again, the typical move is viewed as from science to metaphysics, at least in the larger part of Buchdahl's march through modern philosophy. However, Buchdahl wavers: "The interesting point is indeed that the influence does not just operate from the side of physics on metaphysics, but also in the reverse, for Leibniz's 'metaphysical needs' (if the expression may be permitted) in their turn quite evidently did much to dictate the direction of his

---

4 Thus see Nicholas Rescher: "In mathematical analysis the continuity properties of functions play an important role, and there is little doubt that it was his mathematical studies which suggested to Leibniz the philosophic potentialities of the continuity concept." *The Philosophy of Leibniz* (Englewood Cliffs, N. J., 1967), p. 51, n. 9. Couturat, Russell, Schrecker, rescher, Mittelstrass, Aiton: one must be very bold indeed to oppose the authority of such a group of Leibniz scholars! As will be seen, I do not so much offer opposition as gentle insinuation. It is enough to trouble the sleep of the giants; I cannot expect to match their readings of Leibniz text-by-text and argument-by-argument.

5 Paul Schrecker, "The Unity of Leibniz' Philosophic Thought," in *Leibniz, Monadology and Other Philosophical Essays*, trans. Paul and Anne Schrecker (Indianapolis and New York, 1965), p. xv.

6 Jürgen Mittelstrass and Eric J. Aiton, "Leibniz: Physics, Logic, Metaphysics," in *Dictionary of Scientific Biography* ed. C. C. Gillispie (New York, 1972-75), vol. 8, p. 157.

enquiries into the foundations of physical science, in particular, of dynamics."[7] Nevertheless, as we shall see below, Buchdahl's honest admission does not prevent him from doing the usual thing: regarding the philosophical position as being derived from the scientific one.

Perhaps the easiest way to manage the question of the interaction of philosophical and scientific elements in the work of Leibniz (and a good many other thinkers as well) is forthrightly to admit – what is otherwise true anyway – that for him there can be no legitimate distinction drawn between philosophy and science. The admission does not take us far enough. We still have to think of the total ensemble of ideas as being separable into groups with different stress yield points. Whether there are numbers that are neither zero nor of any definite size; whether *vis viva* can be accurately measured – are questions with something like answers that can can be obtained by calculation and experimentation, construing both of these operations, of course, in their widest form. Contrariwise, whether we are to think of simple substances as monadic or atomic, matter as inert or organic, are questions not easily susceptible to calculational or experimental resolution. Something must therefore be said about the linkage between concepts with high and low stress yield points, and the deliberate blurring of the distinction between philosophy and science will not for this purpose be very helpful.

I am suggesting that what we take to be empirical concepts as parts of typical scientific hypotheses have low stress yield points – someone can always find them to be false; and that metaphysical views have very high stress yield points – sometimes they will yield to no external pressure at all. If the engineering terminology is offensive, think instead in terms of pain thresholds. Empirical hypotheses frequently suffer much pain; even if they don't, they ought to be so constructed that they are liable to suffer pain at any time. Metaphysical views, on the other hand, are very sturdy characters, and for some people, at some times, are regarded as immune to pain.

The readings of Leibniz to which reference has been made seem to me to reverse the stress yield or pain threshold scale. Surely we are no longer in a position to think that empirical statements entail metaphysical ones, or that we can logically derive scientific claims from an overarching and more general worldview. Another prominent consideration is that studies in historiography remain unsuccessful in attempts to give clarity and application to the intuitively satisfying idea of "influence." "Influences" seem to be heavily influenced by the conceptual trappings involved in a given historian's line of approach. What seems to be needed is a different model for understanding the priorities that different thinkers have apparently assigned to the various concepts in stress or in pain in their systems. I suppose that every scientist or philosopher thinks that everything he holds true is true (except for errors of youth, subsequently withdrawn). But a historian might reasonably ask himself the following question when dealing with inference and analogy in the realm of ideas: can I discover the idea, or more likely the set of

[7] Gerd Buchdahl, *Metaphysics and the Philosophy of Science* (Oxford, 1969), p. 407-8.

ideas, that would be the last one that a thinker would give up before going over the brink of conceptual sanity? We like to think of ourselves as operating in a social domain where the etiquette is one of unlimited conceptual largess: we will freely entertain any idea; we will be willing at any time to change our minds. Blind fools! Thinkers weigh their ideas, they give up some, they are never satisfied about others, they remain throughout their careers confused about still others, and there are some they seem unprepared to relinquish considerations of etiquette notwithstanding.

I think that if one applies my suggested historiographic principle in the case of Leibniz the typical stories all become unsatisfactory. It is an unfortunate fact that the case I am about to make for understanding Leibniz is marred by Leibniz' own too accommodating approach to ideas. The Leibnizian character defect must be left unconsidered; his system did have some high stress yield points, or some high pain thresholds. None of these, as I will now proceed to detail, resulted from his very considerable scientific and mathematical achievements. They resulted, rather, from what is often mentioned about Leibniz but never discussed at length, what is known about him but is obscured by the standard influence and analogy accounts: *the entire intellectual career of Leibniz was an attempt to convert what he took to be the insights and consequences of an essentially gnostic Weltanschauung into publicly available scientific knowledge.* An important corollary of this thesis is the proposition that Leibniz apparently did *believe* the basic tenets of a gnostic metaphysics, and that these tenets provided the high stress yield points in his system of thought.[8] Let us see if the appropriate case can be made.

Some of the required circumstantial evidence will be cited later. First, I think we must confront the monad. Perhaps a simple example will set us on our way. We have no trouble thinking of a line as an infinite number of points (mathematical, not physical, points), and as being constructible out of any of its parts (each part of the line contains as many points as does the line, no matter what its length may be). In a perfectly good sense the points are the "parts" of the line, and there are infinitely many such parts. all of this, unfortunately, is true only of a fiction containing two phantoms, point and line. The mathematics of the situation is correct and persuasive (if a bit counterintuitive, as in the case of the theorem in

---

[8] Metaphysics may be seen as a nest of more and more inclusive *motivations*, and the more inclusive the motivation, the more difficult it is to psychoanalyze it away: the motivation to seek maximum clarity about certain problems, and maybe to solve them; the motivation to explain problem solving methods by inclusion in more sweeping, deeper, epistemological or "metaphysical" networks; the motivation to "take up a question" in the first place, to put a certain premium on finding an answer, and to place greatest emphasis upon certain ways of "seeing" or structuring understanding of the world. Clarity motivations led to Leibniz' science; depth motivations produced his "public" metaphysics. The question why he found any of this interesting is only answered by reference to motivations at the third level, the level at which one asks himself: why on earth would Leibniz have thought of replacing physical atoms with spiritual ones? why must everything be thought of as being alive? why are there no gaps in being(after all, we find what appear to be such gaps all the time)? Just try the following exercise: how does one make Leibniz *believable* to late 20th century undergraduate students?

modern algebra that yields the result that we can reconstitute an orange from any of its segments). Consider, however, the case of an actual physical line. The part/whole relationship is characteristic of this line; one thinks that the entire line is decomposable into infinitely many smallest parts. But now the logical problems rush in. If *p* is a proper part of whole *w*, it is ontologically the *same in kind* as *w*. Thus any discovery that *p* is the smallest part of a *w* that is infinitely composite is logically impossible because *p* will possess all of the relevant properties of *w*, one of which is that it will be composite. Any quest for actual material simple substances appears to be logically doomed. Expressed in another way: the mathematics of point/line relationships is physically uninterpretable. The result is a crushing blow to atomism. It also leaves us with only a formal understanding of what it means to be a composite. The question is this: is there not perhaps an alternative to atomism, a strategy that will allow us a more than formal understanding of composites as consisting of "parts" which themselves have no parts? If you begin by thinking that there are wholes, then there are parts, but must every part also be regarded as a whole?

Yes, it must. But only if we limit ourselves to thinking about material wholes. If we turn our attention to what we might call metaphysical wholes, then the outcome is different. For now that which is *simple* in any composite will be without parts, and hence will be subject to none of the accidents characteristic of the composite. These partless simples are of course the monads, and they can be the simple substances of nature only by relinquishing their claim to be *material*. Atomism as a doctrine about matter is false, but atomism as a commitment to simple substances is true. It is false to view matter as extension but it is true to think of everything as being of the same kind – it is not a question of there being no gaps in matter, what is at issue is that there be no gaps in *being*, everything must be of the same kind. It follows from these considerations that what we normally think of as inert, mechanically manipulated matter must be reconceived in a radical way: matter must be of the same kind as the monads or else it is a something else, and the only something else it could be is a nothing, a privation, a ghostly *limit*.[9] Matter cannot be of the same kind as the monads, they are alive and spiritual. It is something else, it is phenomenal, it is constructed out of purely relational, nonsubstantive, ideational spaces and times.[10]

---

9 The literature of gnosticism and Neoplatonism (not to speak of that of much of Christian theology as well) is full of discussions of limits. Why has no one suggested that it is this metaphysical concept of the limit that motivates acceptance of the calculus? Might it be that we are inclined to read the success of that part of mathematics back into the work of Leibniz?

10 In discussion of this paper at Benmiller, Adolph Grünbaum correctly pointed out that my account in this paragraph and the one just before it leaves out of consideration most of the more intricate philosophical problems of space. I concede his point. I do believe that what I have said is, however, a sketch of a successful reading of the first three paragraphs of the *Monadology*. In part I am prompted to read Leibniz in this way because of certain suggestions in Husserl's discussion of the whole/part logic, suggestions too complex to discuss here. See Husserl's analysis of "foundation" and "dependent" and "independent" parts in *Logische Untersuchungen*, vol. 2, part 1, p. 111. *Zur Lehre von den Ganzen*

Thus one can derive the basic features of Leibniz' metaphysics (leaving aside crucial epistemological and methodological refinements) very quickly. Some will take this to mean that I have merely summarized the Leibniz of Couturat and Russell: the *Monadology* is basically a nest of rather simple logical considerations. They are mistaken. Too many choices are open to Leibniz. That his major views fit together logically is not surprising. Schrecker was right in insisting that one of Leibniz' major efforts was that of reducing everything possible to rational forms. Nevertheless the logic is neutral, it imposes no special constraints upon its own interpretation. The same is true of the mathematics. We have just seen how Leibniz' crucially important concept of continuity results from his very conceptualization of the ontology of the monad.[11] The significance of that ontology, I hasten to say, is not given by the discovery of the calculus nor by what Schrecker calls the infinitesimal method of Leibniz. Of course the principle of continuity is basic to the system of Leibniz, but it is basic as a principle of ontology methodologically operative in the production of certain kinds of scientific moves; it is not derived from a mathematics that is neutral with respect to its interpretation.[12]

If it is not the calculus (and also not the physics of forces) that delivers the high stress yield points in the thought of Leibniz, where are these to be located? We have known for a long time, but our quest for a supremely rationalist Leibniz has obscured the vision. Even A. O. Lovejoy could not finally come to admit it; nor Lewis White Beck, whose masterly study of German philosophy gives all the right hints.[13] Paul Schrecker was right: the ultimate source of the *Monadology*, and hence of all of Leibniz' substantive thought, *is* a Platonic myth, except that Leibniz did not regard it as a myth. I invite you first to consider the circumstantial evidence; I will then try to convince you by another of my exercises in revisionist history of

---

und Teilen (Halle a. d. S., 1928), 4th ed.

11 Consider just one of the many texts: *Principles of Nature and of Grace, Based on Reason* (1714), para. 3: "All nature is full [a plenum] ... and because of the plenitude of the world everything is connected ...".

12 If the calculus were liable to metaphysical interpretation only in terms of Leibnizian continuity, what fools Newton and Berkeley must have been!

13 A. O. Lovejoy, *The Great Chain of Being* (Cambridge, Mass., 1936), surely the *locus classicus* for understanding the history of the concept of continuity; Lewis White Beck, *Early German Philosophy: Kant and his Predecessors* (Cambridge, Mass., 1969). Beck suggests a thematic connection between Leibniz and the mystic Nicholas of Cusa (71), notes Leibniz' admiration of the even more mystical Jakob Böhme (156), and in many places is fully aware that Leibniz lived and learned in Germany, a country with a rich and influential mystical and occultist background. Beck understands that the historical move toward rationalism as *Aufklärung* in Germany is rather different from that in Britain or in other parts of continental Europe. The transition from 14th century German piety (the *devotio moderna* of the *Imitation of Christ* and *Theologica Germanica*) to Kant's assertion of Aufklärung as "... man's release from his self-incurred tutelage" is historically unique, and such patterns of difference and diversity are not frequently noted in standard works in the history of philosophy.

philosophy.

Leibniz read everything, wrote to everyone, and filled his works with a dazzling array of citations. It is therefore difficult to know who taught him what about this or that. A number of things that we know about him are nevertheless instructive for my present purposes. He appears to have been a Rosicrucian;[14] while in Nürnburg he was secretary of the lodge. Chemistry was one of his interests during a short period following 1666. The apparent difference between chemistry and alchemy was not at that time decisively made out. In 1566 he published one of his earliest logico-mathematical works, *Dissertation on the Art of Combinations*, in which he cites the *Ars magna* of Raymond Lull, a thirteenth century Franciscan skilled in the number magic and mysticism of the cabala and gnosticism.[15] Leibniz' texts contain numerous references to the mystic Böhme, to Robert Fludd and to Paracelsus. Sometimes he mentions these hermeticists in a very derogatory way. And well he might. The struggle to supply rational form to ancient wisdom does not involve commitment to the excesses of hermeticism, nor does it involve literal acceptance of gnostic "truths". On the thesis here promoted gnosticism supplies Leibniz with a metaphysical motivation. I do not know if the thesis can be strengthened in ways that will make Leibniz unrecognizable even to me.

Consider the following:

> For besides extension and its modifications, there is inherent in matter the very force or power of action which allows *the passage from metaphysics to nature and from material to immaterial things*. This force has its own laws which derive not solely from those absolute and, so to speak, *blind principles of necessity which prevail in mathematics*, but from the principles of perfect reason. Once these higher principles have been established by general research, all natural phenomena can be accounted for and explained mechanically.[16]

---

14 William Shea has pointed out to me that Leibniz could not have been a Rosicrucian because there were no (empirically ascertainable) Rosicrucians. Whenever one of them was about to be detected, he became invisible! Such is the power of the Black Arts. Even contradictions *in being* are tolerated and pragmatically carried out.

15 It may be intriguing to consider: how does the attempt to discover the number of the Beast of the Apocalypse differ in any important way from the attempt to give mathematical grounds for computing that mysterious entity known as an instantaneous rate of change of speed? When, exactly, did alchemy become chemistry and geomatria modern mathematics? Was it not part of the endeavor of the medieval astrologer to have us "sit down and calculate"?

16 Leibniz, *Critical Remarks Concerning the General Part of Descartes' Principles* (1692), "On Article 64", Schrecker & Schrecker, op. cit., 78, emphasis added. That Leibniz' crucial point is methodological rather than substantive is brought out later in the passage:

> At the same time [as one seeks to eliminate resort to substantial forms and souls] one will understand that in former times the Schoolmen erred not so much in dealing with the indivisible forms, but *in applying this theory in places where the question concerned rather the modes and operations of substance and its forms of action, that is, when the question dealt with mechanism*. Nature contains, so to speak, an empire within an empire, or a double government: the

This is a compelling case of Leibniz actually saying the opposite of what his major interpreters proclaim. In what follows he castigates those who have to use substantial forms and souls to sustain their opinions against the *methodological primacy* of mechanical explanation. Apart from the fact that the quotation helps to sustain my thesis in this paper, the reference also contains a helpful anticipation of Kant.

Not only is Leibniz emphasizing the science independent credentials of metaphysics, he is indicating how from that independence results the methodology of positive science in its hardest form: mechanical description and explanation. Given an epistemological repositioning, the quotation summarizes Kant's philosophical system. When the structure of the *a priori* is assured in its application to objects of possible experience, and that structure is systematically secured under "principles of perfect reason", reason in its regulative employment (its methodological employment at the most general level of application) proves that ordinary mechanical explanation is perfectly harmless, indeed, to be preferred because of its simplicity.[17] It is interesting that Leibniz should go on to scold those who do not faithfully shave with Occam's razor. I take it that his point is that metaphysics properly understood and properly managed involves no ontological commitment to supersensible beings as beings recalcitrant to treatment in either rational metaphysics or science proper; instead, *rational metaphysics provides protection for mechanical science as a methodology*. It is a paradox of psychohistory that methodology should, in the thought of Leibniz, provide a bridge between gnostic motivations and fully public and objective natural science. For Kant, the gnostic motivations are replaced by the no less interesting and compelling "demands of reason," but the bridge remains methodological.[18]

> government of reason and the government of necessity [compare Kant: the causality of freedom and the causality of necessity], or the empire of forms and that of material particles. *Just as all is full of souls, all is also full of organic bodies. These two realms remain distinct, each one being governed by its own law* (ibid., p. 79; emphasis added).

It appears that one is free to adopt essentially gnostic themes without thereby committing oneself to a misplaced methodological employment of such topics.

[17] Kant's discussion of the dialectical clash between teleology and mechanism in the *Critique of Judgment* pits one *regulative* principle against another. That, for Kant, the teleology/mechanism issue is one concerning the preferred methodological employment of certain principles as demands of reason, is a point often neglected in accounts of Kant. Some students of Kant wonder why he was even interested in the problem. Because he was a Leibnizian, that's why, and hence the philosophical role of the supersensible (briefly, the hermetical) was of deep concern to him. The "Critique of the Teleological Judgment" is, after all, an essay on the role and status of the supersensible.

[18] Again the line from Leibniz to Kant can be drawn with complete assurance. Leibniz lists the following as ideas only inappropriately applied in mechanical science: the Archeus (an occult vital principle animating and perfecting individual substances), occult fertility and production virtues,

The affinity of Leibniz to the unborn Kant could of course be stressed in many ways. Sufficient unto the present point is the observation that Leibniz would probably have been as upset about the speculative excesses of Swedenborg as was Kant, and for just the same reasons. To find deep explanatory motivation in gnostic themes is not to enter gnostic propositional claims (if there are any such) into competition with public science. It is just not enough in this epistemological competition to trade upon the mysteries and the magic – credentials bearing the mark of some kind of intellectual respectability are required. My thesis, after all, is not that Leibniz was a hermeticist, it is only that his overall position cannot be understood without some accomodation of the nonscientific high stress yield points evident in his work. His attempt, as I read him, was to make secret doctrine public, but this only in the context of a system heavily dependent upon the themes and ecstasies of a gnostic system taken by Leibniz to be the synthesizing force capable of uniting the warring elements in his generous and, for the most part, welcoming intellect. The elements were many: a scholastic and Neoplatonic background, fascination with the new mathematics and physics in whose development he was himself involved, a Lutheran religious commitment coupled with a Calvinist theology and draped with a kind of Catholic intent, and a liberal social conscience more marked in thinkers of the eighteenth Century. Above all, Leibniz had the abiding optimism and confidence in reason (Schrecker is right in this) that made the synthesis seem worthwhile. Lessing wrote that "The great religious truths were not rational when they were revealed, but they were revealed so that they might become so."[19] Leibniz lives somewhere in the vicinity of this possibly insightful observation, taken in an appropriately secularized sense.

My account has so far left undiscussed the important philosophical details of those high stress yield points that I believe Leibniz took over from German mysticism. These points are, if you like, the revelations that underwrite the system; they are the points that provide a motivation for the synthesis, a vital nerve of the methodology that will produce the calculus and the physics of forces. Gnosticism, after all, is not irrational, it is *arational*. It begs for a system and an elaborate cultish ceremonial embodiment. Above all it is what we might call a "secret semantics." In the very wildness of its intricate formalism and its painstaking

---

substantial forms, souls, and "the simple will of the *deus ex machina*" (ibid.). Kant's list includes: circulating humours in organic bodies, a supersensible Being, and thinking spirits without bodies. Kant lists as "things of opinion" (hypotheses appropriate in mechanical explanation) subterranean fire as the cause of earthquakes and volcanoes, rational inhabitants of other planets, and 'the ether of the new physicists." (*Critique of Judgment*, "Methodology of the Teleological Judgment," sections 90-91.) That Kant's impulse to discuss methodological problems at length stems from his acceptance of a "transcendental" strain in Leibniz may be surprising to some; unfortunately for them, this impulse may be the central philosophical anxiety for Kant. [For discussion of Kant's Leibnizian theory of hypothesizing see my "Kant on Hypotheses in the 'Doctrine of Method' and the *Logik*," *Archiv für Geschichte der Philosophie* 44 (1962), especially pp. 197-200.]

[19] From Lessing's *Über die Erziehung des Menschengeschlechtes*; my reference is to Etienne Gilson's citation, *The Spirit of Medieval Philosophy* (New York, 1940), p. 19.

attention to minutiae it endeavors to give a *meaning* to talk about the nature of things. Gnosis may be *secret* knowledge, *but it is thought to be knowledge*. I have referred to Leibniz' mention of "the blind principles of necessity" characteristic of mathematics. The Platonic myth gives sight to those principles. What, then, are the basic elements of that myth?

In my discussion of the nature of the monad I have given expression to one fully essential ingredient – the principle of continuity. I have also suggested that this principle is derived by Leibniz from the concept of the wholeness or fullness of the universe. "*Tout est plein dans la nature*." Nature is a plenum of force; it is alive – all of it is alive. The 3rd century gnostic Valentinus used the term "pleroma" (a term also found in Paul). The Platonism of the myth quickly breaks down. Unlike the eternal and ahistorical Platonic realm of ideas, the pleroma, the nature of Leibniz, the God of Leibniz, has a *history*; it initiates events and in its attempt to give itself in creation or conservation, it does only the best that it can. Gnostic fullness is constrained by the blind necessity of possibility. Choices must be made. (Consider Spinoza: freedom is the recognition of necessity. What necessity? we might reasonably ask. There is no bite, no "dread" in the freedom without the possibilities.) Nor is the technological model of Leibniz' God as supercomputer capable of calculating all possibilities inappropriate; it is just that the model must be understood as being nothing more than a convenient way of displaying by metaphor the metaphysical fact that God or nature knows all, including knowing that the impossible cannot be. The Demiurge and the Pleroma: the two cosmic postures or attitudes of the Gnostic Father.

But what is it, then, that a fullness does? It remains what it is. But to do this, Leibniz thought, requires *effort*. Remember that for him the idea of Hobbes and Spinoza of *conatus* was more basic than the idea of force. The effort of that which is is to remain that which it is: nature, just like any living being, endeavors to maintain itself in a state of equilibrium. Imbalances in the organism, possible failures of light generated by the cessation in application of the network of natural laws, must be regarded as privations dictated by the limits of logic, *and nothing more*. At this point the gnostic theme is paramount. Bent on safeguarding the spirituality of reality at all costs, the gnostics pleaded that matter is a *limit* (interestingly enough), a privation of spirit, a *nothing*, a *darkness*. (What then are we to say about the source of the idea that space and time, the two concepts most basic to talk about physical objects, are only relational, not *really* real?)

These considerations are bound up closely with gnostic ideas about creation as emanation. If reality is full, reality is one. There may be a continuity of objects, events, occasions – but they are all of the same kind. The phenomenal realm of material objects whose behaviour isdescribable mechanically (employing nonsubstantive ideas of space and time) is a kind of Leibnizian equivalent of docetism: the Christ only seems to have a body, the spritual cannot in any sense be physical. The writings of Leibniz are full of this emanationist "logic." In *Discourse on Metaphysics* he says "... it is very obvious that created substances depend on God who conserves them and also produces them continually by a kind of emanation, as we produce our thoughts."[20] This activity of God must of course be exactly reproduced in all that emanates from him, so that all created or conserved beings also possess the power to act, strive and expend effort. In an extraordinary exegetical move, Buchdahl says of all this: "At the back of this there lies the most important of Leibniz' discoveries in mechanics, the principle of *vis viva* ..."[21] It seems to me to be the other way 'round: it is from the acceptance of a concept of nature as a fullness of emanations that Leibniz hits upon the idea of living force. To argue that the "analogy" moves in the opposite direction seems to me just wrong. In fairness to Buchdahl it should be said that he does not in fact *argue* for this direction of the analogy. His discussion does, however, leave exactly the wrong impression.

The theme of emanation is repeated in the *Monadology*. In paragraph 47 we find: "Thus God is the only primitive unit or the only original simple substance, of which all created or derivative monads are the products, born, so to speak, every moment by continual fulgurations from the divinity, and limited by the capacities of creatures, to which limitation is essential."[22] This statement is so crucial for my reading of Leibniz that I need to put heavy emphasis upon it. I think it says the following:

(1) There is only one simple substance, God.
(2) The activity of God is the continual *expression of himself in various forms* (the monads).
(3) Though *of* God, no specific expression of his choice *is* God (although it is, in the full scholastic sense, *like* God), and in this specific sense it is a privation of being, a limitation.

A full commentary on this wonderful passage would produce a kind of handbook for gnostics. All the basic elements are here. All that is is an expression of some divine choice. Each expression is involved in just the same life-history as is the life-history of God. The windowless monads each pursue the same end (driven by *conatus*); each one moves through unreal time in a way that is teleologically equivalent to the movement of any other. Only the mode of expression differs: amethysts "grow," fossils are the "remains" of organisms time-

---

[20] Leibniz, *Discourse on Metaphysics*, trans. Peter Lucas and Leslie Grint (Manchester, 1953), p. 22. The suggestion that our thoughts are always parts of us is typical of Leibniz. Connectedness, continuity, harmony: the themes are repeated over and over again. Later the archrationalist gnostic Hegel will suggest that our possessions are only extensions of our personalities, and we will begin to understand a man by what he buys and builds, not by what he dreams and by his romances. Man is what he makes and what he pays for. From the point of view of psychohistory, is rationalism the basic human pathological state?

[21] Buchdahl, op. cit., p. 395.

[22] *Monadology*, trans. Schrecker and Schrecker, op. cit., p. 155.

frozen in rock – in the living mineral is the dead animal; or, if you prefer the poetry of Leibniz: "... every portion of matter can be conceived as a garden full of plants or as a pond full of fish. But every branch of the plant, every limb of the animal, every drop of its humours, is again such a garden or such a pond."[23]

The idea that the expressions of God's choices are fulgurations is brilliant. The model of light was prominent amongst the gnostics. The building up of the fullness is equivalent to the generation of as much light as possible, keeping in mind the fact that emenations coming later and falling near the boundary of being fade off into darkness (again another limit!). Also, it is suggestive that the consequent end of a lightning bolt is the same in kind as its source, the whole event being only a complex transmission of atmospheric energy. The only difference is that the fulgurated being is less intense, less real, *more limited*, than its source. I can only suggest again that the aeons of Valentinus' pleroma are the natural ancestors of Leibniz' fulgurated monads resulting from the activity of God.

The exposition above is a sketch of a programme for understanding Leibniz. The programme may fail when the history is written in fine detail. Nevertheless, I think that there is a philosophical cohesiveness in what I have tried to portray as the legitimate conceptual and arational background (the high stress yield points) of Leibniz' thought. At the very least, I hope to have been successful in offering an alternative to ways of reading Leibniz *resulting from the subsequent success of science*. The optimism of Leibniz was not born of that success. The physics of forces, the infinitesimals of the calculus, and the subject-predicate logic of Leibniz have all, either individually or severally, been taken to be sources of his major metaphysical ideas. If my programme turns out to be correct, we will simply have to reverse the priorities (and in the bargain, I think, recapture a much more interesting and plausible Leibniz). Certainly Leibniz thought that the physics of *vis viva* is correct, and empirically substantiated. There can also be no doubt that much indeed would have been required to get him to give up the calculus, even though from time to time he expressed doubts about the weird numbers whose existence must be accepted if the calculus is to be believed.

There is, nevertheless, the ugly bit of history that shows him at his very worst while defending his authorship of the calculus. Various ways of reading this historical incident are available. It was not, after all, a question of Leibniz defending the calculus as the source of his philosophical system. It was a case of an irate, although normally overly cooperative, man defending his patent. Think in this connection of the participants in the renaissance "algebra matches." They lied and cheated, stole results from one another, avoided public disclosure of their

---

[23] Ibid., para. 67, p. 159. Not unlike Spinoza: the attributes are after all only ways in which we *conceive* the one substance. It is standard to point out that this fantasy originates from observations made with a good microscope. Perhaps. But compare the section with the rest of the *Monadology* and it will emerge as the poetic expression of a tightly connected vision of the nature of things, a vision so powerfully primal as to give informed sight to the microscope. Galileo is a good case of one who realized that one must be independently motivated to believe what one "sees" by means of an optical instrument.

solutions so that no one else could take the credit for success, were ugly and socially unacceptable during the period of the actual competition, and frequently won their cause by beating the stuffing out of their challengers physically. This kind of behaviour carried forward for a long time in the history of German academe. Even as late as the early part of the twentieth century it was not uncommon for fist fights to break out at public examinations for the doctorate. Indeed, although these fights are no longer likely to cost one an eye or a broken finger, they are still with us in the form of that psychological warfare waged by a dissertation supervisor for the sake of getting his candidate a pass on his oral examination. Was this what Leibniz wanted? On this occasion and forever to be the world's champion calculus discoverer? I am inclined to think so, but the thought is pillowed in the recognition that the infinitesimals of the calculus are quasiformalized fulgurations of the fullness.

Leibniz' acceptance of the Aristotelian containment semantics for the subject-predicate logic is also obviously crucial, but here again the question of the direction of the analogy must be answered in the negative so far as adherence to the standard commentaries is concerned. There is no evidence that the semantics of this logic as understood by Leibniz *produced* his metaphysical ideas. That semantics is a coherent aspect of the doctrines, but it is not determinative. If I am right that the physics of forces was suggested by the organicism of Leibniz' publicly proclaimed systematic gnosticism; if I am right that the infinitesimals are a mathematized expression of the principle of continuity coupled with the basic idea of a monad, then I think I may also be right in suggesting that the fundamental Leibnizian notion that in all true propositions the predicate is fully contained in, or is identical with, the subject is again the statement of a result of having accepted the God-monad-emanation-continuity metaphysics that I have sketched above. Like Plato, Leibniz was not a "systematic" philosopher. What that means, I believe, is that neither of them sat down and wrote one long and boring book in which everything they wanted to say was neatly albeit quite artificially stuck together by the conventions of syntax and semantics. I do think it correct to hold, however, that Leibniz had a system, and that this system was a coherent form of Renaissance gnosticism. To want to make the ancient wisdom public, to want to have it motivate empirical enquiries – even to produce palpable practical results – was a high ambition. Maybe the ambition could only develop its performed self in a time of great religious, philosophical, and social confusion. Maybe only Leibniz, the sometimes successful diplomat, could even dream of activating such an ambition.

At one neglected level of historical consciousness these "maybes" haunt our understanding of great philosophers of the past. Sometimes it is better not to exorcise a ghost; sometimes a historian of philosophy – *especially* a historian of philosophy – should follow the ghost back into those faintly lighted caverns where magic, mystery and experiment incite reason to act on behalf of all of us because alternative forms of primal action become for us too unbearable, too heavy, too inclined to meaninglessness.

CHAPTER 4

RATIONALISM IN MODERN SCIENCE: D'ALEMBERT AND THE *ESPRIT SIMPLISTE*[1]

In the comparison of methods reputed to yield reliable knowledge, it is fashionable to point out that scientific method is the only universally acceptable method of getting knowledge, because scientific claims can always be confronted with the "facts," can be checked against "experience." Every schoolboy knows this is true because he has been taught it many times. It is likely too that he will never forget the other part of the lesson: knowledge gained from experience yields the power to transform the world. As the mountain of technologically engineered objects grows higher and higher the evidence is incontrovertible that science, by sticking to the facts of experience, "produces the goods." And ever since the genuinely remarkable John Dewey fused the Puritan notion of "work-to-beat-hell" with the value hierarchy associated with Yankee ingenuity, we in this country have been content "to buy" (note the pragmatic ring of the phrase) the method which produces the goods. In addition to this widespread contemporary stress on engineering in science, there are prominent philosophies of science that overemphasize the empirical, the inductive, and the pragmatic sides of science.[2]

No one, least of all the historian of science, would deny the therapeutic values of such an approach to knowledge. Yet, if it can be shown that the world is better for having been cleansed of medieval otherworldliness and impractical scholarly preoccupation with trivia, it can also be shown that the account of scientific method fostered by generations of scientists, philosophers, historians and teachers – science as the great consulter of raw facts of experience – is only half the truth about the methodology of modern science. The other half, and to some minds perhaps the essential half, remains partially hidden in the dim recesses of the history of science. This paper will be an attempt to catch some faint glimmering of

---

[1] [Previously published in *Bucknell Review* 8 (1959): 127-139.]

[2] See, for example, the logical empiricism of Carnap, Reichenbach and Bergmann; the varieties of pragmatism and operationalism; and the doctrines of Ayer's positivism, of the American naturalisms that operate under the influence of nineteenth-century British Utilitarianism, and of the Vienna Circle.

an outline of that other half [3] by special reference to non-empirical facets of the thought of the eighteenth-century French mathematician-philosopher, Jean le Rond d'Alembert, whose work is normally grouped with that of his more obviously empiricist contemporaries.

I

Normal characterizations of modern scientific method point to Bacon and Galileo, Newton and Locke, the British empiricists, the French sensationalists and *philosophes*, for confirmation of the view that this method was founded on a rigid empiricism. *Hypotheses non fingo*; replace deduction with induction, harmful certainty with useful probability; out with innate ideas; each idea is an exact copy of some direct impression of the senses; true knowledge comes only through generalizations from things observed in the raw – these are the pillars of scientific method and its attendant empiricist, positivistic epistemology. It is undeniably true that science has been attended by such positions – in Bacon's *Novum Organum*, in Galileo's *Dialogues*, in Newton's *Principia*, in Hume's *Treatise*, in the *Encyclopedie*, and in countless other places in the classical literature of modern science and philosophy. But it is also true that modern scientific procedure – not the one talked about but the one *used* by Galileo, Newton and their followers – was uniquely characterized not by its appeal to *raw* experience but by its introduction of the *controlled experiment.*

The truth is that the experimental discoveries that advanced science beyond the dark ages were not based on random observations of things experienced. In fact, the whole stress of the scientific method of scientists like Galileo and Newton was on keeping variable, subjective human experience out of science. Otherwise, why all the talk about the *objectivity* of science? Stripped to essentials, experiment in Galileo's and in Newton's sense is only a complicated way of putting a question to nature. The question, however, is all-important. No questions, no answers. Observe all you like, but observations do not answer unasked questions. A historian of ideas, Alexandre Koyré, has put this nicely in reference to the work of Galileo:

> It is not "experience," but "experiment," which played . . . a great positive role. Experimentation is the methodical interrogation of nature, an interrogation which presupposes and implies a *language* in which to formulate the questions, and a dictionary which enables us to read and to interpret the answers. . . . Obviously the choice of the language, the decision to employ it, could not be determined by the experience which its use was to make possible. It had to come from other

---

[3] See also E. A. Burtt, *The Metaphysical Foundations of Modern Science* (New York, 1954); A. N. Whitehead, *Science and the Modern World* (New York, 1949); and *Adventures of Ideas* (New York, 1955); Philipp Frank, *Philosophy of Science* (Englewood Cliffs, N. J., 1957), chs. 1, 2, 13, 15; and George de Santillana, "Aspects of Scientific Rationalism in the Nineteenth Century," in *International Encyclopedia of Unified Science*, Vol. II, No. 8 (Chicago, 1950).

sources.[4]

Even Karl Pearson, an otherwise convinced empiricist and positivist in philosophy of science, wrote:

> The reason we find in natural phenomena is surely put there by the only reason of which we have any experience, namely, the human reason. The mind of man in the process of classifying phenomena and formulating natural laws introduces the element of reason into nature. . . .[5]

The point is that experimentation involves an imputation of something *to* nature. Whitehead, always ready with the correct phrase, tells us that "experiment is nothing else than a mode of cooking the facts for the sake of exemplifying the law."[6] Translated into a more common idiom, the suggestion is that we do not get answers to questions we do not ask, and *the nature of the questions determines the nature of the answers we get, if any*. Thus we very nearly obscure the facts entirely when we claim that the difference between ancient-medieval and modern science resides in the fact that Aristotle would not look into Mrs. Aristotle's mouth before he concluded that men had more teeth than women, and that Newton would make an appropriate investigation of Mrs. Newton's mouth. The difference resides more surely in the ways in which Aristotle and Newton put their questions to nature. And the ways, as Koyré suggests, do not come from experience.

If not from experience, from where? From the possession of a scientific credo, a set of commitments more basic than the commitment to experiment or the lack of such a commitment. Aristotle's credo involved a special and very complicated metaphysics. It furnished the motive for the kind of vocabulary he used to ask and answer his scientific questions. Galileo's and Newton's credo also involved a special and very complicated metaphysics. (Perhaps, with these two, a better word would be *presupposition*.)

The credo of early modern science (and it would seem of contemporary science too) is a kind of rationalism – not the metaphysical kind of rationalism (of Descartes, Spinoza, Leibniz and Wolff, or even of Hegel) but the kind of *valuational* rationalism indicated in the quotation above from Pearson. Basically, the rationalist credo involves a question-asking framework in which the notions of simplicity, quantifiablity, measurability and the like are taken as basic. In short, modern science as a rationalist credo means simply the adoption of an essentially *mathematical* vocabulary for use in questioning nature. Mathematics is the soup in which the facts are cooked. In Descartes, and in those scientists who accepted,

---

4 Alexandre Koyré, "Galileo and Plato." *Roots of Scientific Thought*, ed. P. Wiener and A. Noland (New York, 1957). p. 150.

5 Karl Pearson. *The Grammar of Science* (New York, 1957), p. 91.

6 A. N. Whitehead, *Adventures of Ideas* (New York, 1955), p. 94.

explicitly or implicitly, the mathematical rationale of his *Discourse on Method*, the rationalist credo took the form of "*esprit simpliste.*" To discover this underlying rationalist scientific credo, this *esprit simpliste*, beneath the truly staggering modern applications of scientific laws to practical problems is indeed difficult. As George de Santillana states it: ". . . rationalism is never in the spotlight; it is what makes the life of method but is outside method, a persistent form of thought which creates the inner tension necessary for the progress of method. It is for the scientist what the feeling of beauty is for the artist, and hardly more communicable."[7]

II

A clearer view of what I am here calling the rationalist credo of modern science is obtainable by direct reference to a number of great scientists of the seventeenth and eighteenth centuries. The eighteenth-century mathematician-philosopher Jean le Rond d'Alembert is a typical example, though his work is less known than that of Newton or Galileo.

D'Alembert's scientific and philosophical work, on the surface at least, involved what looks like an exact copy of the thorough-going empiricism of David Hume. In outline, this empiricism is reducible to six major heads.

1. D'Alembert subscribed to the sensationalism fashionable in his day and best expressed in the work of Condillac. In the famous "Discours préliminaire" of the *Encyclopédie* he wrote: "It is from our sensations that we derive our ideas." The existence of human sensations is incontestable, and hence sensations must be taken as the basis of all knowledge, an acceptance which replaces the need for innate ideas.

2. Science, according to d'Alembert, was an outgrowth of "human nature," its capacities and its needs. Thus he wrote in the "Discours préliminaire":

> . . . The different ways in which the mind operates on various objects, and the different uses to which those objects may severally be put, constitute the means primarily offered us for distinguishing, one from another, our several kinds of knowledge. Everything is related to our needs of one sort or another, whether it be an absolute necessity, or something that conduces to our comfort and adds to our pleasure, or something we desire from mere habit or caprice.[8]

The primary faculties of the human mind, thought d'Alembert, were memory, reason, and imagination. These yield, in turn, the major divisions of knowledge: history, philosophy or science, and the arts. The classification, rather revealingly, is forthrightly admitted to be that of Bacon.

3. With the *philosophes* generally, following in the tradition established by Montaigne and Bayle, d'Alembert shared an aggressive anti-clericalism and

---

7 de Santillana, *op. cit.*, p. 3.

8 Quoted in Paul Hazard, *European Thought in the 18th Century* (New Haven, 1954), p. 206.

religious skepticism. It was not, of course, a necessary consequence of empiricism that one be a religious skeptic. But in the eighteenth century the fashion was so widespread as to amount almost to a full implication.

4. Since human knowledge arises as the result of personal sensations and is not innate or revealed by a higher authority, so also must ethics be construed on the basis of this sensationalism. The key to ethics is sympathy, a non-rational, empirically based attitude.

5. In the writing and editing of the *Encyclopédie* in collaboration with Diderot, d'Alembert helped to make the philosophy embodied in this great work useful in matters of practical application. To this the large number of items in the *Encyclopédie* dealing with practical, technical household crafts bears ample witness. We find in this professed compendium of all the world's certifiable knowledge numerous definitions and descriptions of mundane things like cookery, the manufacture of needles, the need for closer supervision of midwives, the art of sharpening knives, and the methods of catching fishworms. The point is that it was all a part of the grand Baconian scheme: knowledge is power (for good, he – and the *philosophes* – hoped, rather naively). It was also a part of the new empiricism.

6. Finally, there was the actual point of view which d'Alembert apparently adopted in his own scientific work. In "Système," an article in *Encyclopédie*, d'Alembert mentioned the classification of systems earlier offered by Condillac in his *Traité des systèmes*.[9] According to this classification, there are three types of systems. Of type one (the a priori type), are systems based on general abstract principles for which a priori evidence supposedly exists in confirmation, as exemplified in the metaphysical systems of Descartes, Leibniz, Malebranche, and others. Of type two are those systems based on conjectural hypotheses, which are thus abstract in form but which experience of individual facts was expected to confirm, as exemplified in Descartes's physics. Systems of this kind are now called hypothetico-deductive systems. Of the third type, the only one wholly acceptable to Condillac and to d'Alembert (if we erroneously take some of d'Alembert's remarks too literally), are systems derived purely from facts – the inductive systems of experimentation advocated by Bacon and presumably used by Newton. In what follows, systems of types one, two, and three will be referred to as a priori, hypothetico-deductive, and inductive systems.

The seventeenth-century scientists had bequeathed to eighteenth-century men of thought a hardy and healthy distaste for metaphysical systems of the a priori type. In the "Discourse préliminaire," d'Alembert wrote: "the taste for systems, a taste more appropriate for flattering the imagination than for enlightening the reason is today almost completely banished from sound treatises."[10] And in the article on "Système," written later, he said: "Cartesianism . . . had put the taste for systems very much in vogue . . . [but] today, thanks to Newton, it seems that one

---

9 Jean le Rond d'Alembert, *Oeuvres complètes* (Paris, 1821-22), XV.

10 *Ibid.*, I, 31.

has gotten over that prejudice."[11] In this view, d'Alembert was sharing in the continuing attack on metaphysical appendages to scientific systems of the inductive kind waged by the *philosophes* and by British empirical philosophers like Locke and Hume.

In fact, some of the misunderstanding of modern scientific method might have been engendered by those historians of science and philosophy who thought that in attacking these metaphysically rationalist systems, modern empiricistic scientists were attacking also what I have called "the rationalist credo."

Evidence also exists that d'Alembert was in practice, though not in principle, opposed to systems of the hypothetico-deductive type. In his own scientific work recorded in writings on dynamics and other physical topics,[12] he followed rigorously the physico-mathematical description of objects, and he warned scientists not to prejudge nature by supposing that principles derived from mere calculation (which would yield a hypothetico-deductive system) will be found confirmed in it. "Geometry," he wrote, "which should only obey physics when united to the latter, sometimes commands it."[13] Emile Cailliet has summed the whole matter up as follows: "In the 'Discours préliminaire,' d'Alembert points out . . . that Cartesianism [as an *a priori* system] is contrary to the empiricism of . . . [the eighteenth century]. For him this empiricism is represented by Bacon, Newton, and Locke. The last-named is praised for having renounced a metaphysics of abstractions in order to undertake 'the grand experimental physics of the soul.'"[14]

III

The evidence outlined above can lead to the conclusion that d'Alembert's writings comprise just one more large page in the growing book of modern empirical science and philosophy. But this part of d'Alembert's thought gives us only what I would call the "first half" of the story about his views of scientific method. The other half, the part of his intellectual temperament that led him to want the *esprit simpliste* to play a large part in the life of science, has yet to be uncovered by the historians of science.

In the first place, though d'Alembert was unequivocally opposed to the Cartesian *metaphysics* (Descartes's aprioristic system), he nevertheless sympathized with and employed in his own thinking the basic outlines of Cartesian

---

11 *Ibid.*, XV, 778.

12 For example, *Sur le système du monde*, and *Essai d'une nouvelle théorie de la resistance des fluides*.

13 Quoted in de Santillana, *op. cit.*, p. 8; citing *Essai d'une nouvelle théorie de la resistance des fluides*.

14 Emile Cailliet, *La tradition littéraire des idéologues* (Philadelphia, 1943), p. 101. Author's translation.

method. In this he was not in full accord with Condillac. Thus he wrote: ". . . properly speaking, the principal merit of the physicist is to have the *esprit de système*, and never to make anything of it."[15] The suggestion here is that though a sense of simplicity and systematization should govern the scientist's use of method, simplicity and systematization as methodological postulates should not be generalized into metaphysical propositions regarded as a priori truths. Aram Vartanian has pointed to another relevant passage in d'Alembert's work. "Although Descartes's physics itself began to be discarded, as d'Alembert remarked, some thirty years before the first volume of the *Encyclopedie* appeared in 1751, the scientific methodology that had inspired it was to fare more fortunately."[16]

There are two ways in which this sympathy for Cartesian method influenced the thought of d'Alembert. First, Descartes's method gave him the desire to reduce all of knowledge to a simple set of unifying principles, a desire clearly set forth in the "Discours préliminaire." Second, in d'Alembert's own scientific work, the high place accorded to mathematical modes of thought, a place accepted by d'Alembert uncritically and almost passionately ("For me mathematics has been a mistress!"[17]) was influenced by Descartes's own mathematical predispositions.

The first evidence of Cartesian influence on d'Alembert (it is to be found similarly in Diderot) is found in the rationale of the *Encyclopédie*:

> It is perhaps true to say that there is hardly a science or art in which, if necessary, the most limited intelligence could not be instructed by the use of good logic; for there are very few sciences and arts whose laws or propositions cannot be reduced to simple notations, and arranged in so immediately perceptible an order that the chain would not be broken at any point.[18]

The Cartesian flavor of this remark is unmistakable. The idea of reduction to simple principles introduced by the *Discourse on Method* is further employed by d'Alembert in his definition of science itself. For him, science is ". . . the art of reducing, so far as possible, a great number of phenomena to a single one which could be considered the source of all."[19] Pursuing further the analogy of general

---

15 D'Alembert, *Oeuvres*, I, 77.

16 Aram Vartanian, *Diderot and Descartes, A Study of Scientific Naturalism in the Enlightenment* (Princeton, 1953), p. 139. The reference is to *Oeuvres*, I, 73. Vartanian's study shows impressively the extent to which the thought of the *philosophes* derives from the *esprit simpliste* of Cartesian method. The present paper owes much to Vartanian's extensive knowledge of eighteenth-century thought and his perceptive disclosure of its largely rationalist sources.

17 Quoted in Joseph Bertrand, *D'Alembert* (Paris, 1889), p. 32.

18 D'Alembert, *Oeuvres*, I, 34.

19 *Ibid.* I, pp. 27-28.

scientific method to mathematical analysis, d'Alembert went on to claim that "the universe, for the person able to contemplate it from a single point of reference, would be only a unique fact and one great truth."[20]

This set of remarks on reduction to simple principles of all more disparate and complicated scientific statements stands in sharp contrast to d'Alembert's professed empiricism. A rigid empiricism delights in diversity and plurality. To wish to reduce the many to the one, to find unity in diversity, requires the rationalist credo, the *esprit simpliste*, which took shape in d'Alembert's remark that the principal merit of physics is its *esprit de système*.

It also requires an unwavering confidence in mathematical language as the vocabulary for interrogating nature. And d'Alembert took mathematics to be preeminently the valid form of universal rationality. It is only on the basis of such a confidence that d'Alembert could write the following statement, which again sharply contrasts with his empiricism.

> The great metaphysical problem has been put recently: are the laws of nature necessary or contingent? To settle our ideas on this question, we must first reduce it to the only reasonable meaning it can possibly have . . . viz., whether the laws of equilibrium and motion that we observe in nature are different from those that matter would have followed, if abandoned to itself. . . . Hence this is the way the scientist should follow: first he should try to discover through reason alone which would be the laws of mechanics in matter abandoned to itself; then he should investigate empirically what are really such laws in the universe. If the two sets of laws be different, he should conclude that the laws of mechanics, such as those yielded by experiment, are of contingent truth, since they appear to spring from a particular and express decision of the Supreme Being; if, on the other side, the laws yielded by experiment agree with those that could be deduced by logic alone, he shall conclude that those laws are of necessary truth: which does not mean that the Creator could not have established a wholly different set of laws, but that he did not hold it right to establish other laws than those which resulted from the very existence of matter.[21]

One of the things that stands out in this passage is d'Alembert's posing of the question in the framework of the distinction between necessary and contingent truths, a framework provided by Galileo, Descartes, Leibniz, and Spinoza. The contingent laws, by d'Alembert's own admission, are those established only observationally by empirical investigation. Now an empiricist, Hume, for the very best example, would likely admit that these are the only laws we have. From this admission we go either in the direction of skepticism or of practice. We replace science by "a deeper faith" as the romanticists and Bergson sought to do, or we appeal to the fact that after all science "works," and the man on the street whose car runs and whose aspirin relieves him of pain will settle for just that. But d'Alembert, the apparent empiricist schooled in the skeptical tradition, takes neither of these

---

20 *Ibid.*, I, 33.

21 *Traité de dynamique* (1758), p. xxiv; quoted in de Santillana, *op. cit.*, p. 10.

directions, because for him the laws of nature *are* necessary, since finally the uniformities of ordering observed in nature miraculously *agree* with the laws deduced from the very nature of matter (abandoned to itself) by purely logical, i.e., mathematical means. Of course what happens is that the experimental vocabulary of mathematics prejudices the outcome. What we get is a science of the hypothetico-deductive type in Condillac's listing: not an empty listing of observations, but a nature ordered according to the questions we put to it.

What is particularly instructive about d'Alembert's stated scientific method (yielding a hypothetico-deductive system) is its *direction*. We are not urged to attempt the formulation of laws that agree with the regularities we observe. Quite the contrary, the direction we are to follow is the peculiar and surely paradoxical line of discovering whether or not the observations agree with the laws of mechanics that are deduced from the "very existence of matter!" The problem is to get empirical reality to agree with mathematical structure, and not to get mathematical structure to conform to reality as observed. This is not, moreover, merely a matter of wording. The relation involved is asymmetrical. We want observed facts to agree with mathematical deductions, and not conversely. For d'Alembert and most modern scientists, it would make no sense to talk about theory conforming to fact; and if we find facts which do conform to theory, it does not logically follow that therefore theory conforms to facts. To see the import of this distinction, let us substitute the more contemporary term "confirmation" for d'Alembert's less precise "agreement." The relation of confirmation is asymmetrical with respect to scientific systems. For if a fact x confirms a theory y, it makes no sense to say that therefore y in turn confirms x; y *might organize, systematize* or correlate x with other facts m, n, o, etc., but it does not confirm x. In common language, a fact is a fact because it is (a fact), implying that facts stand as they are with respect to evidence, and need no independent confirmation.

The theory of scientific method in d'Alembert's characterization of laws of nature is thus the exact opposite of Bacon's inductive method. This strongly suggests that modern science was not born because Kepler, Galileo, Newton and others applied Bacon's non-Aristotelian logic; it was born because they used successfully a generalized form of exactly Aristotle's logic – not the syllogistic, but the general method of deductive inference in which Aristotle and the scholastics placed such confidence – expressed in the more precise and more fruitful forms of mathematical procedure developed by the scientists themselves. It was their confidence in the forms of this more comprehensive logic of mathematics (the general outlines of which are given in Descartes's *Discourse on Method*) which provided the basic proposition in the credo of early modern science.

IV

What emerges from this brief look at the two halves of d'Alembert's thought on scientific method, though it may look like a paradoxical result, is only a special form of the general scientific method, *plus* the credo motivating that method, which was in fact employed by early modern scientists. The method itself turns out on close inspection to be the hypothetico-deductive method, added to which are the experimental procedures used in confirming the hypotheses. Mathematics, in any of the special forms used, provides the vocabulary which is used in translating the experimental results *into* the system of deductively arranged hypotheses. The resulting system, in spite of Condillac's (or Hume's) protests, is of the hypothetico-deductive sort.

Thus the two halves of d'Alembert's philosophico-scientific personality are joined. The empiricism is there, and with it goes a sympathy for purely inductive and experimental systems. The function of the empiricism, a function in which modern empiricism is truly triumphant, is to combat aprioristic metaphysical systems which require pseudo-evidence for confirmation and which do not yield the power for technical control over nature. The rationalism is there also, not as a metaphysics, but as the *esprit simpliste* or the *esprit de système*.

The rationalism is at the same time a motive force and an extra-scientific passion. It shows the scientist as an artist rather than as a tabulator, collector, or mere calculator. The life of science breathes in it, and science as a distinctively human creation stands forth because of it. D'Alembert put the point nicely: mathematics was his mistress. It would be difficult to find a more apt expression of the rationalist credo. Its expression in such terms is only possible because d'Alembert, trusting only in part his Baconianism, took deductive logic (in its mathematically more precise and elegant forms) as the paradigmatic form of system-building. Through the integration obtained by mathematical reduction, he and the Encyclopedists generally wished to build a brave new world. They did it too, and the success is still with us, though perhaps we misunderstand it. We are more likely to understand it if we look to Descartes rather than Bacon, to the complex deductive systems characteristic of modern physics rather than to Galileo's experiments with balls rolling down inclined planes.

The war against traditional metaphysics (systems of the a priori type) has been won. It is doubtful that science would gain anything by a return to the unchecked kind of metaphysical speculation characteristic of the seventeenth century. Nevertheless, contemporary philosophers of science (especially positivists and operationalists) and many scientists themselves would better help us to understand science if, rather than stressing empiricism and the pragmatic method of "problem solving," they emphasized that scientific strength is in scientific *systems* and that one of the obvious and necessary bases of the system is the creativity of the scientist. Gustav Bergmann, a logical positivist, tells us that "scientists observe what happens. Whenever they can, they manipulate things so that they may observe what happens under certain circumstances. This helps them to discover laws of nature. Having discovered some, they try to combine them into theories."[22] All of this is true. But Bergmann's account, in the light of the rich understanding of the scientist found in the work of d'Alembert, is anemic indeed. Without the *esprit simpliste* or the *esprit de système* the scientist is an imitator or

---

22 Gustav Bergmann, *Philosophy of Science* (Madison, Wis., 1957), p. 3.

an appendage to the instruments over which he thinks he has control. But he has no share, as I think d'Alembert has, in what Whitehead called "the adventures of ideas." The scientific adventure surely involves, as d'Alembert saw, explanation as well as description, imaginative construction and projection of hypotheses as well as manipulation of instruments, and mathematical elegance as well as technical application.

CHAPTER 5

HUME'S SCEPTICISM[1]

There are today two rather widely accepted but divergent ways of viewing Hume's philosophy. The first sees Hume as having advanced a radical scepticism which devastated the very possibility of scientific knowledge. The other regards Hume as having introduced a healthy scepticism prompting philosophers and scientists to abandon the rash claims of modern rationalists and to embrace humbly the fact that non-trivial knowledge is after all only probable. The first of these views of Hume stems from Kant. The second appears to prevail amongst various kinds of contemporary philosophers, and stems from logical positivism or later logical empiricism.

The formulation of Kant's view is given clearly in the first *Critique* and even more clearly in the *Prolegomena*. In the first *Critique* Kant refers to the "*censorship* of reason," Hume's inference of "the nullity of all pretensions of reason to advance beyond the empirical,"[2] and his general view is that Hume "...gave himself over entirely to *scepticism*, having, as he believed, discovered that what had hitherto been regarded as reason was but all-prevalent illusion infecting our faculty of knowledge."[3] In the *Prolegomena* Kant refers this scepticism to Hume's presumed use of the analytic-synthetic distinction as Kant himself used it:

> Hume being prompted to cast his eye over the whole field of *a priori* cognitions in which human understanding claims such mighty possessions (a calling he felt worthy of a philosopher) heedlessly severed from it a whole, and indeed its most valuable, province, namely, pure mathematics; for he imagined its nature, or, so to speak, the state constitution of this empire depended on totally different principles, namely, on the law of contradiction alone; and although he did not divide judgments in this manner formally and universally as I have done here, what he said was equivalent to this: that mathematics contains only analytical, but, metaphysics synthetical, *a priori* propositions. In this, however, he was

---

1 [Previously published in *Journal of the History of Ideas* 20 (1959): 413-419.] This paper was read at the Oct. 27, 1957 meeting of the Creighton Club at Cazenovia Junior College. Cf. my dissertation, *Husserl's Criticism of Hume's Theory of Knowledge* (1957), written under the supervision of Professor Paul Schrecker of the University of Pennsylvania.

2 Immanuel Kant, *Critique of Pure Reason* [1781, 1787], A 760, B 788; trans. Norman Kemp Smith (London, 1950), 607.

3 *Ibid.*, 128.

> greatly mistaken, and the mistake had a decidedly injurious effect upon his whole conception. But for this, he would have extended his question concerning the origin of our synthetical judgments far beyond the metaphysical concept of causality and included in it the possibility of mathematics *a priori* also, for this latter he must have assumed to be equally synthetical. And then he could not have based his metaphysical propositions on mere experience without subjecting the axioms of mathematics equally to experience, a thing he was far too acute to do.[4]

Representing the positivist or logical empiricist interpretation of Hume's philosophy, Reichenbach writes:

> He [Hume] arrives at the result that all knowledge is either analytic or derived from experience: mathematics and logic are analytic, all synthetic knowledge is derived from experience. By 'derived' he means not only that concepts have their origin in sense perception but also that sense perception is the source of the validity of all nonanalytic knowledge. The addition to knowledge supplied by the mind, therefore, is of an empty nature.[5]

Crucial in both the Kantian and the logical empiricist readings of Hume is the ascription to his thinking of the analytic-synthetic distinction, or, at the very least, both views interpret Hume's thinking as having involved something very nearly like this distinction. Here, however, the similarity between the two views ends. According to Kant, Hume separated mathematics as analytic from all other knowledge as synthetic, where all other knowledge means metaphysics, natural science, and everyday factual convictions. Because, for Hume, under this interpretation, analyticity stood as the only mark of genuinely certain knowledge, his scepticism resulted from the discovery that all other claims to knowledge have only the character of synthetic, or not absolutely certain, judgments. Thus, according to Kant's interpretation, the discovery of Hume which led to the devastation of natural science and hence to philosophical scepticism amounted to finding that not all knowledge is analytic. Here Kant stepped into the breach with his synthetic *a priori* in an attempt to revise Hume and thus save us from his damaging scepticism.

The logical empiricist, on the other hand, always unwilling to accept the synthetic *a priori*, simply decides that Hume is the genuine parent of his (the logical empiricist's) apparently necessary distinction between analytic *a priori* and synthetic *a posteriori* statements, these kinds of statements counting as the only allowable epistemically significant ones.

Leaving aside these differences in interpretation, prompted by differences in epistemological approaches, the important thing about the ways in which both

---

4 Immanuel Kant, *Prolegomena to Any Future Metaphysics* [1783], (Prussian Academy Edition, 1911), 272; Lewis W. Beck translation (New York, 1951).

5 Hans Reichenbach, *The Rise of Scientific Philosophy* (Berkeley and Los Angeles, 1956), 86.

Kant and logical empiricists read Hume is that both parties apparently think that Hume's philosophy established or involved the (not explicitly made) distinction between analytic and synthetic judgments or statements. Now if this claim is true, then, according to Kantian and normal contemporary usage, we must regard Hume as having been interested in the *logical characterization of types of judgments or statements*. If Hume was interested in such a logical problem, certain apparently overlooked problems of an intriguing character arise. For Hume's writings, though often they give the external appearance of having been involved in logical analysis, are now fairly generally regarded as treatises in *psychological* philosophy. That is to say, in crucial places in the *Treatise* and the *Enquiry*, Hume gave psychological explanations even though the problems involved called for answers requiring logical analysis. Such are Hume's psychological answers to the problems of causal inference, general ideas, probability logic, and others.[6]

So it seems essential to ask, though I know of only one place where it has been asked,[7] whether it makes any genuine sense to read Hume as having concerned himself with a logical characterization of types of statements such as results in the distinction between analytic and synthetic statements.

It appears true to say of Hume that his philosophical thinking retained one timeworn and tenacious vestige of rationalism – he was interested, as is obvious from his remarks in the Introduction to the *Treatise* – in finding a system of knowledge which, certain in itself, would justify the findings of the various special sciences. What apparently motivated Hume, then, was the desire to characterize *certain* knowledge, and of course such a characterization required him to study problems of evidence, problems of justification. Thus Hume wanted to know what kinds of evidence we have for sciences accepted as yielding reliable knowledge, and whether these kinds of evidence will also support sciences of a more dubious character, dubious because their methods require projections into the future. His conclusion was that only the sciences of quantity and number (arithmetic and algebra) can be "pronounced...objects of knowledge and demonstration."[8] And his reason for this conclusion was that these sciences deal only with relations between ideas, or, put in non-Humean language, these sciences have evidence for their assertions which does not refer to experience, or to matters of existence.

Now this reading of Hume, which I think is generally faithful, would

---

6 This point is conclusively established in Edmund Husserl's *Erste Philosophie, Erster Teil, Kritische Ideengeschechte* (The Hague, 1956); and *Logische Untersuchengen, Vol. I, Prolegomena zur Reinen Logik* (Halle A. D. S., 1928, 4th edition); and in Norman Kemp Smith's *The Philosophy of David Hume* (London, 1941). Cf. Charles S. Peirce, "The Laws of Nature and Hume's Argument against Miracles," *Values in a Universe of Chance, Selected Writings of Charles S. Peirce*, ed. Philip P. Weiner (Garden City, 1958), 289.

7 Husserl, *Erste Philosophie*, 350ff.

8 David Hume, *Enquiry Concerning the Human Understanding* [1748], ed. L. A. Selby-Bigge (Oxford, 1955, 2nd ed.), 163.

incline one toward the view that Hume was really interested in seeing how far the realm of *a priori* evidence extends. His discovery, viewed in this way, would then amount to finding that evidence of an *a priori* kind does not support assertions in the sciences of experience, because the evidence for these sciences is wholly *a posteriori*. If it makes sense to read Hume in this way then what really issues from his philosophy is not the logical distinction between analytic and synthetic, but the methodological or epistemological distinction between *a priori* and *a posteriori* kinds of evidence. Thus Hume was not interested in marking off precisely types of statements, but in discovering the kinds of evidence permissible for any kinds of statements whatever.

These remarks, however, do not meet the views of Kant and the logical empiricists head on. What is needed in addition is a direct confrontation of the question: did Hume's philosophy in fact involve something like the distinction between analytic and synthetic? Since giving an answer to this question might prove impossible if we endeavored first to find out what the terms "analytic" and "synthetic" *really* mean, I propose to use them in their Kantian senses, a proposal which seems justified when it is kept in mind that the issue under discussion in this paper results from Kant's interpretation of Hume. Kant's familiar definition of an analytic judgment is that any judgment in which the predicate is covertly contained in the subject is analytic. Or, to use his words in a variant definition, "Analytic judgments (affirmative) are therefore those in which the connection of the predicate with the subject is thought through identity...."[9] Synthetic judgments, for Kant, are those whose predicates are not covertly contained in the subjects, or in shorter form, synthetic judgments are all those which are non-analytic. So the question to be investigated should now read: is Hume's distinction between "relations of ideas" and "matters of fact" the same as Kant's distinction between "analytic" and "synthetic" judgments? It is to be noted that if the answer to this question is negative, then one is left with the interesting job of finding another, non-Kantian, explanation for Hume's scepticism, since Kant's explanation depends on equating the sets of distinctions.

There are two rather strong reasons which might incline one to accept Kant's equation of the distinctions. Both of them have to do with Hume's characterization of "relations of ideas." He states rather emphatically that the evidence for relations of ideas is simply that any statements expressing such a relation cannot be denied without contradiction. This likewise was Kant's criterion for analytic statements. Second, in the *Enquiry* Hume states that "Propositions of this kind are discoverable by the mere operation of thought, without dependence on what is anywhere existent in the universe."[10] Now both of these characteristics of relations of ideas seem also to be characteristics of Kant's analytic judgments. But from this evidence we cannot jump to the conclusion that the equation is

---

9 Kant, *Critique of Pure Reason*, A 7,B 10; *op. cit.*, 48.

10 Hume, *op. cit.*, 25.

warranted, since other matters remain for consideration.

As Edmund Husserl has pointed out in a note to one of his lectures in *Erste Philosophie*,[11] if we are to regard relations of ideas as analytic, in Kant's sense, then we must also regard each statement of a relation of ideas as tautological. Yet Hume's writings do not seem to support such an interpretation. Surely, for him, all identity statements are statements of relations of ideas, but the stronger statement, "all relations of ideas are identities," does not seem to be true. For Hume all that is required for a bit of knowledge to be knowledge of relations of ideas is that we be able to know that it is genuinely knowledge without reference to matters of existence and that we cannot deny this knowledge without contradiction. But the identity of subject and predicate in a judgment of relation between ideas is nowhere asserted by Hume, and for a very sound reason. Hume's basic conception of knowledge is that all knowledge arises as the result of certain psychological principles of association, together with beliefs resulting from habitual observations and actions. This is true even where our knowledge is knowledge of abstract or general ideas, since Hume was at great pains to show that general ideas are actually individual, and that they function generally by representation (an extensional notion) which is possible only because the mind has the power or disposition to recall resembling individuals and to arrange them in similarity-classes.

This being so, all knowledge for Hume depends on psychological synthesis, on habit, and on the ability to recall past happenings. Thus knowledge of relations of ideas cannot be merely tautological, since for Hume all knowledge is basically empirical and hence non-trivial. Even identity, where Hume talks about it, is given an extensional treatment depending upon his conception of general ideas and his conception of similarity. Of course, to say that all knowledge results from psychological syntheses is not also to say that all knowledge statements have the logical character of empirical statements. But the point is that for Hume giving a psychological explanation amounted to the same thing as giving a logical characterization of types of judgments, and for this reason his statements of relations of ideas cannot rightly be called analytic, or identity judgments, in Kant's sense.

Another way of seeing that Hume's philosophy must be understood as saying that all statements are empirical (or synthetic, if you like) is by reference to his notorious distinction of ideas and impressions. Every idea, according to this view, is an exact replica of an impression, an immediately given, separately inspectible, incorrigible datum of sensation. The clarity and distinctness of an idea, to borrow a phrase from the rationalists, is always suspect from Hume's point of view, a suspicion which he voices by referring to ideas as "less lively" than impressions.

Here the point is that if every idea is an exact image of a sense impression (an experience) and thus has the same visual or sensory characteristics as a datum

---

11 Husserl, *op. cit.*, 351-52.

of presentational experience, and if all relatings of ideas take place as the result of principles of psychological association on the basis of habit, no statement of a relation between ideas can be anything but empirical, or synthetic, employing the Kantian terminology. The only difference between an empirical statement of a relation between ideas and an equally empirical statement of a matter of fact is that the latter makes direct reference to the existence of something, while the former does not posit the existence of anything, but, in Hume's phrase, depends "solely upon ideas."[12] Beyond this, the further difference between the two, which Hume's analysis brings out, is that the evidence for empirical statements depending "solely upon ideas" is never the same as the evidence for empirical statements positing existence in a spatio-temporal context.

Again, what I am suggesting is that where Hume states that relations of ideas[13] depend "solely upon ideas" and "are discoverable by the mere operation of thought,"[14] all that he means is that no phenomenal impressions must be called directly into use in the verification and understanding of these relations. And this, I would further suggest, is exactly what we must regard as Hume's implied meaning for what others call "*a priori.*" To depend "solely upon ideas" is to be understandable and confirmable independently of experience (in the sense of direct experience of impressions); thus to depend "solely upon ideas" and hence not at all upon impressions is to be *a priori*.

The above remarks tend, I think, to support two conclusions. First, Kant was mistaken in thinking that Hume's terms "relations of ideas" and "matters of fact" are directly equivalent to his terms "analytic" and "synthetic." Since it has been shown that for Hume all judgments are properly speaking empirical, there seems no further point in considering him one of the parents of the contemporary logical empiricist's analytic-synthetic distinction. Second, my arguments support the view mentioned earlier that Hume was not specifically interested in characterizing logical types of statements whatsoever; and this means that Hume was interested in seeing whether different types of empirical statements can all be said to depend "solely upon ideas." Or, if the Kantian terminology is preferred, Hume was interested in finding out whether all knowledge, like the exact and certain knowledge of arithmetic and algebra, is synthetic and *a priori* (of course in Hume's sense of "*a priori*" as explained above, not in Kant's sense).

Now this gives us a clue to the reason for Hume's scepticism. Hume, like any good philosopher with a rationalist background to draw from, was interested in locating and describing those sciences which are exact and certain. He found

---

12 David Hume, *A Treatise of Human Nature* [1739], ed. L. A. Selby-Bigge (Oxford, 1951), 70.

13 Specifically, relations of resemblance, contrariety, degrees in quality, and proportions in quantity or number – the four philosophical relations which are for Hume "objects of knowledge and certainty." *Ibid.*

14 Hume, *Enquiry*, 25.

these sciences to be algebra and arithmetic. And he found that the evidence for these sciences resides solely in ideas and in their relations of resemblance, contrariety, degrees of quality, and proportions in quantity or number. The conclusion is inevitable: for Hume, *a priori* evidence (dependence "solely upon ideas"), in the form of non-contradiction, is the mark of genuine knowledge, hence of genuine science. As Husserl, to whom I am indebted for this point, puts it: "Rationality or, what amounts to the same thing, *a priority* is the character of science."[15] Surely this is a rationalist thesis, but just as surely it was Hume's thesis as well. And if "*a priori*," as argued above, is to mean "residing solely in ideas," that is, depending in no way upon actual impressions (which pertain to matters of existence), only those sciences which in no crucial way depend upon actual impressions, i.e., *a priori* science, or sciences for which there is wholly *a priori* evidence, can be said to be exact and certain, can be said to possess the "clarity and distinctness" requisite for reliable knowledge.

Of course, Hume discovered that the evidence for natural sciences or factual sciences is not *a priori*, does not reside solely in ideas and their relations, but is *a posteriori* – is evidence derived from actually given phenomenal impressions the apprehension of which involves the positing of existence. And since Hume was working (perhaps not consciously) within the framework of the rationalist thesis which Husserl mentions, his scepticism resulted from this discovery that the evidence for factual scientific conclusions (by inductive inferences, for example) is not *a priori*, lying only in ideas, but rather *a posteriori*, depending upon actual impressions.

Again turning to Husserl's suggestive comments on Hume's philosophy and its implications, Hume's scepticism may be seen as having arisen as a result of the incompatibility of the rationalist thesis that all genuine knowledge has *a priori* evidence and the empiricist theory of knowledge which Hume used to investigate the extension of this thesis.[16] For here lies the real difficulty which Hume encountered. His wish was to found all sciences upon the science of the *facts* of human nature, a science which Hume insists can only have an *a posteriori* (wholly experiential in the narrow sense of impression-experience) justification. Thus sciences, even the exact and certain ones, are to be based upon a foundational science which is itself not exact and certain. This circularity in the outcome of his system, this contradiction generated by trying to realize rationalist aims by empiricist means – this was the source of Hume's scepticism, and at the same time

---

15 Husserl, *op. cit.*, 351.

16 Husserl argues this point in both works mentioned in note 5 above. He also re-makes the point in the posthumously published *Erfahrung und Urteil, Untersuchungen zur Genealogie der Logik* (Hamburg, 1954, 2nd ed.), 472-75.

the source of a good deal of later philosophy, including Kant's.[17]

[17] A word ought to be said about the reasons for Kant's mistaken views on the nature of Hume's sceptical conclusions. Kant missed the point I have argued above, and which Husserl was the first to suggest, because, as Husserl also points out (*Erste Philsophie*, 352), Kant's knowledge of Hume was based almost entirely on a reading of the *Essays*. Thus he did not know enough about the general plan which Hume set for himself in the *Treatise*, nor did he see the fuller working out of Hume's system, with its great reliance on the impression-idea distinction and on analysis of issues not related to the causality problem. Considering only the *Enquiry*, one might make out a case, as Kant thought that he had, for the equation of his statement-type distinction with Hume's evidence-type distinction. Except that even here Kant was shortsighted, since the examples from mathematics which Hume gives in this work (25) as representing statements of relations of ideas Kant would himself have regarded as synthetic and *a priori*. The fact that the set of Kantian mathematical (synthetic *a priori*) statements and the set of Hume's statements of relations of ideas have the same extension ought to have prompted Kant to consider the question discussed in this paper. For further evidence that Kant had only slight knowledge of the *Treatise*, if any at all, see Norman Kemp Smith, *A Commentary on Kant's Critique of Pure Reason* (New York, 1950), xxv-xxix, 156, 161ff., 592-601; and J. M. Keynes and P. Sraffa, *An Abstract of a Treatise of Human Nature 1740* (Cambridge at the University Press, 1938), vii note.

CHAPTER 6

HUSSERL'S CRITIQUE OF HUME'S NOTION OF DISTINCTIONS OF REASON[1]

I

It is commonly overlooked that Hume's section on abstract ideas in the Treatise[2] actually proposes answers to two quite different, though intimately related, questions.[3] Having endorsed Berkeley's nominalism, with its conclusion that generality only arises because particular ideas function representatively, i.e., stand for or represent (denotatively) other particular ideas of the "same sort," Hume's first task was to explain how particular ideas function "beyond their nature" "as if (they) were universal." Hume's answer to this question was, of course, psychological. He ascribed to general names the capacity to stimulate in the imagination the disposition to recall the other resembling particulars. And though there are difficulties in the notion of disposition which trouble both theoretical psychologists and philosophers of science, some recent commentators on Hume's theory of general ideas have praised this dispositional theory.[4]

Hume's second problem in the section on abstract ideas – and this is what is generally overlooked – is the following. Every particular idea is to be presumed

---

1 [Previously published in *Philosophy and Phenomenological Research* 20 (1959): 213-221.] This paper is a revision of Part iii of the author's doctoral dissertation, *Husserl's Criticisms of Hume's Theory of Knowledge* (1957), written under the supervision of Professor Paul Schrecker of the University of Pennsylvania.

2 David Hume, *A Treatise of Human Nature*, edited by L. A. Selby-Bigge, Oxford, The Clarendon Press, 1951, Pt. I, Bk. I, Sect. Vii, pp. 17-25.

3 I am indebted to Edmund Husserl for this insight. See *Logische Untersuchungen*, Band II, Teil 1, pp. 188-89, fourth edition, Halle A. D. S.: Max Niemeyer, 1928.

4 See R. I. Aaron, *The Theory of Universals*, Oxford, The Clarendon Press, 1952, pp. 75-85; and Andrew Ushenko, "Hume's Theory of General Ideas," *The Review of Metaphysics*, Vol. IX, No. 2, Dec. 1955, p. 236 ff.

to be ordered under many groups of similarity,[5] yet each time it functions representatively it stands for ideas falling into only one such group. The white cube resembles both the black cube and the white sphere, yet the general name "white" recalls only the image of the white sphere, together with other images of white objects, and not the image of the black cube. Hume's second problem was then the problem of showing how it is that these similarity groups are not confused in thinking, making generalization more an obfuscation than a clarification and simplification of thinking, as it is usually thought to be. How, when I employ the name "white," and my mental association at the time of employment is the image of a white cube, do I prevent my mind from calling up associations with different colored cubes, instead of different white objects? Hume's second task was to answer this kind of question without admitting universals, which had already been outlawed in the dispositional part of his representative theory of general meaning.

If Hume's theory of abstract ideas is seen as involving these two related questions, and it seldom is, then there is no difficulty at all in understanding why Hume devoted the last couple of pages of this section of the *Treatise* to the scholastic idea of distinctions of reason, because it was this very idea which Hume thought answered his second problem. It is worthy of note that Norman Kemp Smith, an otherwise very astute and perceptive student of Hume, does not read Hume's theory as dealing separately with the two mentioned problems, so his explanation of why Hume introduced the distinctions of reason is simply that Hume could not get along without universals after all, and the closing pages of the section are simply his confused way of admitting this.[6]

II

What Hume wrote about the *distinctio rationis* needs only summary treatment here.[7] Suppose we were to compare a white sphere with a black sphere and then with a white cube. The comparison leads us to note two different similarities. Through repeated comparisons of this sort, objects come to be arranged for us into similarity groups, and we learn through "habit" to "view them in different aspects." Thus, for example, when we consider the color white, we consider its resemblance to the white cube, together with all other remembered white objects, accompanying our ideas "with a kind of reflection, of which custom renders us, in a great measure, insensible." In this insensible reflection a white

---

5 Thus Hume writes: "'Tis certain that the mind wou'd never have dream'd of distinguishing a figure from the body figur'd, as being in reality neither distinguishable, nor different, nor separable; did it not observe, that *even in this simplicity there might be contain'd many different resemblances and relations*." *Op. cit*., p. 25. Italics mine.

6 Norman Kemp Smith, *The Philosophy of David Hume*, London, Macmillan and Co., 1941, p.264 ff.

7 This account follows Hume, *op. cit*., p. 25.

sphere appears and there appears conjointly a resemblance with respect to the color, thus ordering the white sphere in the similarity group of color. In this rather special sense this kind of reflection can be said to "abstract" the distinguished property.

Thus, it is by means of a habitual comparing of simple objects, a habit which eventually renders us insensible to the kind of reflection which makes the comparing possible, that objects are thought together with respect to just certain of their similarities, and it is to be presumed that this kind of reflection, of which we are now largely, if not wholly, insensible, keeps us from confusing different similarity groups. In this way Hume's theory of the distinctions of reason supposedly answers the second question which his inquiry into abstract ideas poses. Here, as in the case of Hume's first problem, his answer is psychological, and this, as we shall see, in great measure accounts for the unsatisfactoriness of his theory of general ideas as a logical characterization of this type of meaning.

III

I should like now to turn to Edmund Husserl's analysis of Hume's use of the *distinctio rationis* as given in *Logische Untersuchungen*,[8] an analysis which shows that Hume's attempt at an answer to his second question is greatly unsatisfactory.

In Husserl's estimation, Hume's *distinctions of reason* doctrine is stated ambiguously, and can be given both a "moderate" and a "radical" interpretation. On the "moderate" reading of this passage in the *Treatise* Hume's view is that every concrete phenomenal object (impression or simple idea) is absolutely simple in the sense that its characteristics are inseparable from it. If this was Hume's intended view, then a *distinction of reason* is merely a "mental pointing" and generalization is to be attributed to the psychological power of *attention*. This, of course, was a part of the theory of Berkeley. Though Husserl criticizes this "moderate" interpretation at length in his consideration of Berkeley's theory of attention,[9] only the "radical" interpretation will be dealt with in this paper.

The second possible way of viewing Hume's theory of *distinctions of reason* (the "radical" interpretation) involves the quite paradoxical implication that different, mutually inseparable characteristics of presented contents (color, form, and the like), which we believe are apprehended as parts present in the contents, are not really in them at all. There is only one kind of real parts – those parts which can also appear separately by themselves, or Hume's "simple ideas." The "radical" interpretation further suggests that abstract partial contents, for example, color qualities, are in a sense mere fictions, since though they can be considered by

---

8 Husserl, *op. cit.*, p. 192 ff.

9 *Ibid.*, p. 137 ff. Cf. Smith, *op. cit.*, p. 266. Smith evidently thought this was Hume's only intended view, and that it covertly admits the generality of abstract ideas, thus destroying Hume's case for the particularity of abstract ideas.

themselves (by means of *distinctions of reason*) they cannot be or be observed by themselves. Thus color is not in the colored content, nor is form in the content formed; there are only those similarity groups into which phenomenal objects are grouped, and certain habits which attend the intuition of phenomena – habits as unconscious dispositions (imperceptible psychical occurrences stimulated by the intuition).

On this "radical" view of Hume's meaning, he has paid a great price indeed in order to complete his representative theory of general ideas by means of the *distinctio rationis*. For if the only real parts in any presented content are the simples, all supposed partial contents of a presentation, which ordinarily is viewed as a unified whole, are not really parts of it at all. Thus if I am thinking about blue whales and my thinking is accompanied by the intentional image of a blue whale with tiny yellow eyes and an enormous gusher of water coming out of its spout, I am not really entitled to say that the blue of the whale and the yellow of the eyes and the form of the spouting water are parts of the presented image at all. And all of this seems to run directly counter to the commonsense and Platonic adage that "I perceive what I perceive when I perceive it."

Remember also that this paradox is generated by Hume's introduction of the *distinctions of reason* part of his theory in an attempt to answer the question: how is it that an object which is in many ways similar to other things is apprehended in a given experience as bearing *just this* similarity to other objects? In order to show the general inferiority of Hume's answer, Husserl proposes this argument:

> If the abstract contents which, in the concrete intuition, correspond to the absolute characteristics are themselves nothing, then the connective and relational contents in the intuition of a formally unified totality are *a fortiori* nothing. Obviously the problem of the *distinctio rationis* and the principle of its solution are identical for all abstract contents. They are identical, therefore, for the relational and connective contents and for the absolute contents. The question how the apparent presentation of the color in the colored object, or the distinction of the color from this object, comes about, cannot be answered, therefore, through a reference to a presented similarity between this colored object and other colored objects. Indeed, if such an explanation is followed up consistently, this presentation would lead back to the presentation of a similarity relation between this similarity and other similarities (in the example of color: to a group of similarities of similarities as they obtain between colored objects). And to this similarity the principle of explanation would again have to be applied, and so on.[10]

In short, Hume's implied conclusion that where we believe we perceive, say, a quality white, we are actually perceiving or otherwise presenting only a certain similarity between the appearing object and other objects, involves his *distinctio rationis* explanation in an infinite regress of the vicious kind.

---

[10] Husserl, *op. cit.*, p. 196. I am grateful to Professor Schrecker for several suggestions pertaining to my translation of this passage.

Thus Husserl's argument leads us to conclude that the theory of the *distinction of reason*, based on a special use of the similarity relation, does not count as a valid addition to Hume's theory of general meaning, because the explanation employing the notion of the *distinction of reason* leads to a vicious infinite regress. Caution ought to be observed in judging Husserl's argument. On the face of it, it looks very much like the usual, now defunct, rationalist rejoinder to the resemblance theory of general ideas as first advocated by Berkeley and Hume.[11] The orthodox argument, which has, I think, been adequately refuted by Aaron and Price,[12] among others, contends that the resemblance theory really covertly assumes what it wishes to dispense with – the very idea of resemblance on which the theory rests must itself be regarded as a universal, as a common relational quality in some Platonic sense: if this is not done, then the theory involves the vicious infinite regress. Now although Husserl accepted, mistakenly, this orthodox argument,[13] I do not think that the argument quoted above involves even the implied conclusion that Hume's use of the *distinctio rationis* covertly assumes the existence of universals, an existence obviated by Hume's representative view of general meaning.

IV

In order to see that Husserl's criticism of Hume's explanation by *distinction of reason* does not involve the ordinary form of the infinite regress argument, consider the line of thought proposed by R. I. Aaron. He claims that the infinite regress argument (in its orthodox form) has been assumed to be a fatal objection to the resemblance theory of universals, but continues with the assertion:

> ... I should like to question the assumption. For admitting the infinite regress, does this make the argument invalid? What the resemblance theory needs by way of presupposition is that we should be able to recognize a resemblance when we see one. Now we do see that the resemblance between *a* and *b* resembles the resemblance between *x* and *y*. And we see this without having to attempt the impossible task of observing an infinite series of resemblances. Supposing we have a case where *a* is true if *b* is true, and *b* is true if *c* is true, and *c* is true if *d* is true, and so on, *ad infinitum*. Then admittedly we could not know that *a* was true. But our present case is a different one. The regress is there, but we *can* know the resemblance in question without observing the infinity of

---

[11] In general, the resemblance theory simply states that, contrary to both realism and conceptualism, there are no universals, but generalization is possible because we are able to note resemblances between objects, which may or may not be viewed as possessing identical common qualities. See Aaron, *op. cit.*, p. 151 ff. General words refer, denotatively, to the resembling objects in the given group. Forms of this theory were advanced by both Berkeley and Hume.

[12] Aaron, *op. cit.*, p. 153; H. H. Price, *Thinking and Experience*, Cambridge, Harvard University Press, 1953, pp. 23-26.

[13] Husserl, *op. cit.*, p. 115.

> resemblances. Consequently the argument does not refute the Resemblance theory.[14]

I think it will have to be admitted that Aaron's suggestion counts against the orthodox infinite regress argument as directed against the normal formulation of the resemblance theory of general ideas. But it does not likewise count against Husserl's argument against Hume's special use of resemblance in his theory of the *distinctio rationis*. The reason for this can best be seen by considering how Husserl's argumeny differs from the orthodox infinite regress objection, which states that to argue away universals on the basis of the similarity relation is covertly to allow at least one universal, namely, resemblance itself.

Husserl's argument tells us that if we allow that there is a resemblance between green and blue and between violet and aquamarine, then there is *at least one distinction* which we are able to make with respect to qualities of objects which are *ex hypothesi* not to be distinguished. *The resemblance itself* is distinguishable in the presentation of the two resembling cases. But Hume's position does not entitle one to make even this distinction, and so again the apparently distinct resemblance – the case where the resemblance of green and blue resembles the resemblance of violet and aquamarine – must be accounted for by a new *distinction of reason*, and so on to infinity. More and more such pseudo distinctions must follow as necessary consequences of Hume's view precisely because without them he has admitted what he does not wish to admit – that distinct aspects of complex ideas are distinguishable. Thus Aaron's refutation of the orthodox infinite regress objection does not hold for Husserl's objection to Hume's view when he (Aaron) asserts that it is true that we can know the resemblance in question; for we could only know this in every case – in cases of simple resemblance as well as cases of resemblances of resemblances – by virtue of a *distinction of reason*, which is after all artificial, since it is no real distinction at all.

It is thus the *distinctio rationis* itself which makes the vicious infinite regress a consequence of Hume's doctrine, and the really vital point is that this doctrine is unable to explain how in a given experience I single out just this resemblance between phenomenal objects of a given kind, rather than one of the many other resemblances of which they are susceptible. In order to begin to explain this I must admit that presented contents have really distinct parts.

V

It follows from this line of approach that Husserl's form of the infinite regress argument as directed against Hume is not the ordinary form of this argument – that to allow resemblance is covertly to allow universals – and hence Aaron's objection does not apply. The validity of Husserl's argument must therefore be judged independently of the demonstrated invalidity of the orthodox

---

14 Aaron, *op. cit.*, p. 153.

infinite regress objection to the resemblance theory.

Husserl's really substantial point is that Hume's *explanation* of abstraction by means of the *distinctio rationis* is an inadequate explanation, because it goes to infinity. Now I think this is surely a valid point, especially if we strip the situation of all extraneous references to universals. Thus Hume is read as offering the *distinctio rationis* as an explanatory hypothesis and all that needs asking is the question of whether it really does explain. If the *distinctio rationis* hypothesis is to count as a correct explanatory hypothesis, we must be able to say what kind of evidence would count against it. We know what this evidence would be, of course; any discovery of a genuine partial content, any part aside from a "concrete" part, a "simple," would count against the hypothesis. But clearly such a discovery will never arise in Hume's attempted explanation, since if we claim that in his explanation a real part is allowed, namely, the resemblance itself, he will push back his explanation one step further by means of a new *distinction of reason* depending upon a new resemblance of higher order. Thus although we know what kind of evidence would invalidate his view, Hume is really telling us that we shall never find any evidence of this sort, and his proposed explanation can be patched up at each stage of the game by the introduction of an artificial distinction. It is in this sense that Hume's proposed explanation leads to infinity, and it is in this sense that his explanation is no explanation at all. In order to save the theory and thus not to make any real distinctions between parts and their supposed wholes, Hume is forced into the uncomfortable position of having to say, and surely there is no evidence for sayig it, that no evidence will ever be found that will invalidate his closed hypothesis. And this is exactly why his explanation is, as Husserl argues, inadequate, and hence does not validly extend his resemblance theory of general ideas.

VI

Husserl's argument then, if my reading of it is correct, throws light both on Hume's reasons for including the often overlooked and seldom understood pages on *distinctions of reason* and on the general character of his theory of general ideas. It appears, first, that Hume needed the *distinctions of reason* in order to have his theory insure the fact that different groups of resemblances which objects under consideration share are not confused in thinking. If his theory of general ideas had led to the view that resemblances cannot somehow be disentangled from the many groups into which they fall, the theory would very clearly have violated an empirical fact which Locke noticed, and Hume endorsed, namely, the employment of general ideas leads to greater economy in thinking.[15] Thinking would not be done economically, if resemblance groups were confused in it; in fact, it would not be thinking at all, but the kind of lack of thinking which characterizes the mental

[15] John Locke, *An Essay Concerning Human Understanding*, edited by A. S. Pringle-Pattison, Oxford, The Clarendon Press, 1934, Bk. III, Ch. 3, Sect. 20, p. 237; and Hume, *op. cit.*, p. 20.

activities of those who find it impossible to generalize at all, or who generalize haphazardly and often incorrectly – for example, primitive peoples, very young children and those suffering from various kinds of brain injury, particularly those causing amnesia.

Husserl's argument, therefore, and this leads to the second major conclusion of this paper, refutes more than he thought that it did. For it not only refutes Hume's doctrine of the *distinctio rationis*, it also refutes the representative theory of general meaning which Hume thought needed reinforcement by means of the *distinctio rationis* addition. Husserl's reading of Hume's theory of general ideas indicates very clearly that whether Hume knew it or not, his theory was in fact an empirical hypothesis of the psychological sort. In full, the hypothesis is as follows: general words have the disposition to stimulate the imagination to recall particulars which resemble that one particular which is associated (as an image) with the general word; *and* the different kinds of resemblance classes into which objects are grouped are kept separate in the mind by means of an "insensible reflection" which Hume calls a *distinction of reason*. Husserl's argument shows that, as an empirical psychological hypothesis, the second part of Hume's full hypothesis is closed, i. e., there is no possibility of finding evidence which would refute it. but since, as I have indicated, Hume's hypothesis needs confirmation of both conjuncts in order to satisfy the empirical fact that general ideas simplify and thus render economical our thinking, the full hypothesis of Hume is refuted by Husserl's argument, though I do not think that Husserl himself was aware of this more general implication of his argument, though he was aware that there are other arguments which count against the first conjunct of Hume's hypothesis.

The weaknesses of Hume's theory of general meaning which the implications of Husserl's argument detect are, it seems fair to remark, simply further signs of the general unsatisfactoriness of psychologism, the major defects of which were pointed out by Husserl, Bradley and Frege. In Hume's case, his desire was to reduce all sciences to the science of the facts (the psychological facts) of human nature. In this sense for Hume all science was psychology (in his peculiar sense of psychology as the science of human nature). And once we realize that his attempt to characterize general meaning was not an attempt at logical or philosophical analysis, but rather an attempt at armchair psychology, Husserl's argument establishes the invalidity on logical grounds of the hypothesis which Hume apparently thought proved on the basis of experience.

CHAPTER 7

THE ROLE OF ARATIONAL FACTORS IN INTERPRETIVE HISTORY: THE CASE OF KANT AND ESP[1]

The Challenge of the Strong Programme. The strong programme in sociology of knowledge is a *programme*; it is a format for research and a philosophical manifesto. If it did not provide a reasoned context for doing research it would be of no great interest to sociologists; if it professed no epistemology it would be of little interest to philosophers. Philosophers of science (in growing numbers, it seems) find a challenge in the tenets of the strong programme. That challenge has largely to do with the fact that the strong programme offers an alternative epistemology in the name of scientific sociology, and many philosophers find it an alternative sadly lacking in philosophical justification.

Much attention has been paid to the strong programme's symmetry principle, and to the problems thought to arise from applying that principle to causes or conditions of scientists' beliefs that are not rational. The symmetry principle flies in the face of the Mertonian arationality assumption discussed sympathetically and at length by Laudan in *Progress and Its Problems*. The principle challenges, among much else, the longstanding philosophical contention that reasons can be causes, and that in seeking to explain choices of beliefs, we should look for reasons for those beliefs, arguments in favor of those beliefs, before we give up the game and turn to social conditions – or worse, psychological ones – as explanations. In our heart of hearts we all know that social and psychological conditioning has an important effect upon what we believe, what we conclude, and what we accept. Feyerabend and Kuhn would not have such a profound hold on us if we did not know this. But philosophers are committed to the inherent or consequential rationality of science (and philosophy as well), and so for many of us acceptance of the symmetry principle is impossible. It may be that Leibniz accepted the calculus as a reliable mathematical instrument because of his gnostic upbringing, but this does not explain why the calculus was accepted by Leibniz as good science; to explain this we need to find the reasons for Leibniz' acceptance, the good arguments that he thought could be put forward on behalf of that

---

[1] [Previously published in J. R. Brown (ed.), *Scientific Rationality: The Sociological Turn* (Dordrecht: Reidel, 1984), pp. 209-247.]

acceptance.

Philosophers' reluctance to accept the strong programme, and especially its symmetry principle, seems to me to stem in part from philosophers' deep commitment to the proposition that science/philosophy really is the paradigm of rational activity, and to the following propositions that this commitment implies:

1. The epistemology of Western Science is the preferred epistemology.

2. The methodology of Western Science is the preferred methodology.

3. The rationality of Western Science is the only form of rationality that requires philosophical justification, and it is the only form of rationality that can be so justified.

4. Western Science is essentially philosophical: it investigates on the basis of shared, public methods; in so far as it is possible it infers canonically; it concludes on the basis of reasoned argument; it accepts on the basis of reasoned faith (in induction).

5. However one makes out the case, Western Science is progressive, it is cognitively successful – it works.

These commitments as stated will not stand unchallenged. All of us would at once rush in with qualifications, nice distinctions, editorial suggestions – as academics are accustomed to do, we would form a committee and take votes. But that won't do. Science and philosophy ought not to be done by committees. Taking a vote is an abdication of the responsibility to undertake the hard work of reaching a universally acceptable rational conclusion. Science/philosophy does not give notices of motion, it issues communications, publishes results. The communications may fail to communicate, and the results may be wrong, but we need to find this out, and to make suitable corrections, *by means of science*, by means of rational argument, including experimentation and perfection of instrumentation.

Advocates of the strong programme can probably accept all that I have said so far. accepting my characterization of the commitments of a rational Western Man, they would simply go on to point out that all they seek is freedom (from philosophers' carping) to engage in the kind of science we admire. Essentially what they seek is license to investigate science itself scientifically. They know (at least Bloor and Barnes seem to know) that science is a body of knowledge as well as a set of social and political activities, and for this reason they also know that studying science scientifically involves appeal to many sciences, among them sociology and psychology. If sociology and psychology of science were content to limit investigation of what we might call the cultural background of scientific activity, the external social and psychological trappings of scientific work, there is no negative philosophical response that could make sense.

However, the strong programme is not an extension of Mertonian diffidence. Instead, it promises to investigate *the internal theoretical workings* of science scientifically (sociologically). To do this it will abrogate the arationality assumption and seek direct sociological explanations of why scientists accept the results that they do. To many philosophers, this means that explanations of scientific inferring and concluding will *effectively bypass* two central disciplines: theory of rationality (taken to be philosophically essential in characterizing the nature of scientific theory); and intellectual history in the form of history of science (taken to be the central ancillary discipline of philosophy of science, providing, as it does, the data of science as a cognitive activity). It is this threatened bypass operation that some philosophers perceive to be the unkind cut.

Philosophers are not troubled by sociologists who study scientists as members of institutional groups, as people who testify before appropriation committees, who perform on television. It is only when it is alleged that these activities of scientists have any bearing on the cogency of their theoretical or experimental results that philosophers bestir themselves. As Laudan has shown, it is the status of *cognitive* sociology that is at issue, it is the sociological study of scientists' beliefs.

Arational Factors and Utilities in Historical Explanation.[2] I have myself engaged in studies of interpretive history of ideas, and often my interests have led me to what some regard as the seamier side of scientific or philosophical belief. For example, without hesitation I attributed Leibniz' acceptance of the strange theory of the monads to his gnostic and mystical background, a background that most of us would identify as irrational, or certainly arational. The preferred mode of studying Leibniz or any other figure in the history of science/philosophy, at least amongst historians of philosophy, involves bringing the thought of Leibniz up to today's standards of rationality. The point of the exercise is to show how much in the thought of Leibniz is of lasting value because it maps onto a present presumed-to-be-correct theory of science. I have no objection to studies in this mode. They are important and illuminating. Using our present understanding of logic to help capture what Leibniz was doing when he invented the prototype of our logic is indispensable. Using our present understanding of physics to help capture Leibniz' concept of living force is likewise indispensable. To dispense with current forms of rational paradigms is much like dispensing with the activity of intellectual history altogether. For, among other things, that history has responsibility for giving an account of the continuities in scientific thought. And we cannot know why we are where we are unless we relate where we are to where others have been.

But there is more to consider. Leibniz did write the German documents, was for a time secretary of a Rosicrucian lodge, did talk about angels. Strong programme advocates would, I suppose, want to make these features of Leibniz'

---

[2] The material in this essay dealing with Kant and spiritualism is more fully discussed in Chapter III of my book, *Kant and the Double Government Methodology: Supersensibility and Method in Kant's Philosophy of Science* (Dordrecht: Reidel, 1984).

activities part of the explanation – the sociological explanation – of his scientific beliefs. At the very least they would want to hold that explanation of these beliefs is not different in kind from explanation of Leibniz' "correct" scientific beliefs. Laudan, I suppose, would want to say that they are part of the metaphysical constitution of a research tradition, and that in the long run they turned out to be nonprogressive features of that tradition. Certainly they do not constitute reasons for Leibniz' acceptance of his progressive scientific beliefs. But on neither of these views is the seamy side of Leibniz eliminated. It should not be eliminated. It is there for all to see. It must be dealt with. The only question is how? Allow me to move to another example in an attempt to show the role I think the intellectual historian can play in clarifying the issues, if not in resolving the dispute. The scheduled bypass surgery must alarm the historian of ideas as much as it alarms the philosopher.

Before I discuss the example let me anticipate the moral of my story – I mean to dull the edge of the scalpel before it cuts very deep. I believe that what are sometimes taken to be social and other external factors do play a role in explaining scientific beliefs, but that this role is not performed by these often arational factors taken as universal ingredient in explanations, or as instances of such generalizations. Scientists undertake their studies for reasons that are better seen as expected utilities. Scientific explaining is, as is often remarked, goal directed. A scientist does his work with certain ends in view (as Dewey would put it), and often, if not always, he will accept the results of a given piece of work because he also believes that the work has achieved those ends in view. Maybe all he seeks is the truth, but epistemic utilities are still utilities.

Accounting for the satisfaction of goals as reasons for accepting scientific results is usually done in ways that see utilities as internal features of scientific induction. For Bayesians, the utilities may be expressed as prior probabilities and may refer informally to expectations of success based on kinds of past success. Wesley Salmon has suggested that the history of science functions in Bayesian inference in something like this way in his provocative "Bayes's Theorem and the History of Science."[3] I will seek to show quite informally that intellectual history appeals to expected fulfillments of ends in view as explanations of accepted beliefs in something like this Bayesian way.

What I am suggesting is this. One form of explanation (Gutting's Type II explanation) has a sociological generalization as its major premise:

(i) Anyone in circumstance C believes p. (*Sociological generalization*).
(ii) S is in circumstance C.
(iii) Therefore, S believes p.

---

[3] In *Historical and Philosophical Perspectives of Science*, ed. Roger Stuewer, *Minnesota Studies in the Philosophy of Science*, V (Minneapolis 1970). A quite different, but promising approach to these problems is to be found in James Brown's doctoral dissertation, *Models of Rationality and the History of Science* (University of Western Ontario 1981).

The form of explanation I am now advancing replaces the sociological generalization with an expressed or implicit end in view, a *meta-reason* for undertaking the research in the first place. This gives us a modification of Gutting's Type I explanation:

(i) S believes p'.
(ii) S reasons in manner R (*for meta-reason M*) and believes the results of his reasoning.
(iii) Given p', reasoning in manner R (*for meta-reason M*) results in p.
(iv) Therefore, S believes p.

I am aware that meta-reasoned reasoning is often self-serving, and that its employment can validate deductively formulated explanations in unacceptably circular ways. (A reasoner's meta-reason for accepting reasoning in manner R might be that he will believe anything that follows in manner R from p'.) My suggestion is simply this: sometimes, at least, a reason for engaging in research of a given kind (and employing a methodology of a given kind) is not a social or psychological or other form of *external cause*, but is a reason as an end in view (maybe only the end of discovering the truth), that is effectively internal to the task of research. When the intellectual historian lights upon such a meta-reason, as he often does, he has discovered an important and relevant *internal reason* for the scientist's belief. And this is true even if the meta-reason is, as it often is, an *areason*, that is, something that the scientist accepts for *whatever reason*.[4]

Much of what is involved in the suggestions I am making has to do, I think, with one of the deep problems inhering in the strong programme, and one not often enough remarked. Advocates of the strong programme have the wrong conception of philosophy! I will not push the point, but it does seem to me that Bloor accepts as his model of philosophizing just that kind of sleight of hand informal analysis of language that is most distant from the work of scientific reasoning. I think one can agree that if philosophy is seen as the kind of thing done by Austin and Flew, then scientists, sociologists included, should put as much distance btween themselves and philosophy as possible. But there is another kind of philosophy, one that has as ancestors those who would not have recognized the

---

[4] Maybe what I am here calling meta-reasons are Gutting's "basic beliefs." I'm not sure about this. I do think there are such creatures as basic beliefs, and that these can be described following identification. If all that the strong programme advocate needs is the assimilation of basic beliefs into my kind of Type I meta-reason explanation, then that programme begins to me to look quite innocuous, because then the scientific questionableness of Type II explanations need never arise, and the absence in sociology of knowledge of any established law-candidate generalizations need cause no embarrassment. I am not suggesting that what I am calling meta-reasons cannot themselves be argued for. Often they are revealed as suppressed premises in the explanation, but are themselves conclusions of independent reasoning.

difference between philosophy and science – Descartes, Newton, Leibniz, Kant – and as more recent practitioners Whewell, Mill, Peirce, Dewey. This kind of philosophy – call it what you will – sees philosophy as at least continuous with science, and it is the view of philosophy as basically not distinct from science that underwrites the kind of analysis I am giving here.

A Case Study: Kant and the Paranormal. Now my case study in the history of ideas, one that I hope will confirm the results of the analysis given above, and will seek again to emphasize Laudan's surely correct observation that "*the application of cognitive sociology to historical cases must await the prior results of the application of the methods of intellectual history to those cases.*"[5]

My example is the case of Immanuel Kant's attitudes toward the doctrines of Immanuel Swedenborg and in general toward alleged cases of extra-sensory perception. The case should interest historians of science because in discussing Swedenborg Kant introduces an encapsulated theory of matter, remarks upon a solution to the mind/body problem, including intriguing references to physiological explanation of mental events, and continues the development of the "double government" methodology of Leibniz. Philosophers should be interested in the case because of the connection between Kant's discussion of Swedenborg and Kant's own later ethical theory. Up until now, there has been little interest shown by either philosophers or historians of science in Kant's views on ESP, except by those who want to show that Kant's references to a spiritual world entail a theory of spirits that is derived, but not acknowledged by Kant to have been so derived, from Swedenborg.[6]

In 1766 Kant published what is generally regarded by Kant scholars as his most remarkable, because atypical, book: *Träume eines Geistersehers erläutert durch die Träume der Metaphysik.* The 'Geisterseher' is Swedenborg. Unlike all of Kant's other works, this book is written in a partly playful style; it teases and insinuates; it is contemptuous. C. D. Broad thought that it abounds "in elephantine badinage." Broad thought – incorrectly – that there is only one other place where Kant discusses or refers to Swedenborg, the letter to Miss Charlotte von Knoblach of 1763.[7] We now know that Kant made a regular practice of referring to Swedenborg in lectures on rational psychology. There are also references in two

---

5 *Progress and Its Problems* (Berkeley & London 1977), p. 208. I am afraid that Laudan will not be completely satisfied with what I am up to: showing that the arational functions in rational explanations.

6 The exception is the very full and competent study by C. D. Broad, "Immanuel Kant and Psychical Research," *Religion, Philosophy and Psychical Research* (New York 1953), pp. 116-155. I will rely heavily on Broad's clear analysis of the case, but will have to update the information he had available to him.

7 In a postscript to his essay Broad adds to the literature, introducing reference to Kant's *Lectures on Rational Psychology*, in which Kant mentions Swedenborg with approval. The date of the letter to Miss von Knobloch is disputed; I think Broad's dating of 1763 is likely correct.

later works, and two Reflexionen.[8] I have also been able to find three more letters, two to Moses Mendelssohn dealing with *Träume*,[9] and a late (1790) letter to Borowski, in which Kant deals with the *Schwärmerei*, the enthusiastic or ecstatic dream state, brought about by too much reading.[10]

Kant's letter to Charlotte von Knobloch is evidently a reply to a letter from this lady to Kant.[11] Kant details three now well-known cases in which it had been reported that Swedenborg experienced supernormal cognition or acheived extrasensory perception. Kant had learned – the details need not trouble us here – that Swedenborg was reputed to have had a clairvoyant conversation with the dead brother of the Swedish queen, during which conversation Swedenborg learned a secret known only to the queen; that Swedenborg had "discovered" a lost receipt for payment to a goldsmith in the secret compartment of a dead man's bureau; and that Swedenborg, while in Göteborg – 459 kilometers from Stockholm – had "seen" a fire that was stopped just three doors from his own house. Kant had sent a letter to Swedenborg inquiring about these alleged incidents which evidently Swedenborg received. But he did not reply. One of Kant's English merchant friends had been dispatched to discuss the incidents with Swedenborg, which he dutifully did. He reported that Swedenborg had received Kant's letter, would reply in a book soon to appear, and that the best testimony the Englishman could find attested to the truth of the stories reported to Kant. Kant adds that he has received other stories of Swedenborg's extensive communications with spirits in the supersensory world.

---

8 That Kant made regular reference to Swedenborg in lectures on metaphysics is now amply evidenced by published transcripts of notes of Kant's students. See *Metaphysik L1*, Kant's *Schriften*, 28.1, pp. 298-99; *Metaphysik L2*, *Schriften*, 28.2.1, p. 593; *Metaphysik Herder*, *Schriften* 28.1, pp. 113-14 (compare *Nachtrage Herder*, *Schriften*, 28.1, pp. 857-88); *Fragment einer späteren Rationaltheologie nach Baumbach*, *Schriften*, 28.2.2, pp. 1324-25. Additional references are in *Strife of the Faculties*, Ak. Ed., Vol. 7, p. 46; and in *Anthropology*, Ak. Ed., Vol. 7, p. 191. See also *Reflexionen* 1486 and 5026, the latter containing this fascinating claim: "That on which the plausibility of a metaphysical hypothesis (Swedenborg) rests. On a putative intellectual intuition on analogy with the sensible one". Can it be that Swedenborg is the source of a doctrine (there is intellectual intuition) that Kant will finally put aside decisively in *Critique of Pure Reason*?

9 February 7, 1766, and April 8, 1766. The earlier letter is not extant, but is quoted in *Kants Werke*, ed. Rosenkranz u. Schubert, XI, I (1842). The second letter is translated in Arnulf Zweig, *Kant, Philosophical Correspondence* 1759-99 (Chicago 1967), pp. 54-57.

10 Kant had purchased, for £7 sterling, the eight large quarto volume set of Swedenborg's *Arcana Coelestia*, a work subjecting every word in the Biblical books of *Genesis* and *Exodus* to a symbolical interpretation. Broad says "It may fairly be described as one of the most boring and absurd productions of any human pen." Kant read the volumes. It is tempting to think that he took the Schwärmerei to have been generated by this kind of reading! Twenty-four years after writing the *Träume* Kant was still worrying about spiritualist ecstacies.

11 It is a long letter, running to nearly 2,000 words. Ludwig Ernst Borowski reprinted the German original of the letter in his *Darstellung des Lebens und Charakters Immanuel Kants* (Konigsberg 1804), pp. 211-25.

Kant begins his letter to Charlotte von Knobloch by expressing charitable agnosticism concerning the reported feats of mystical communication attributed to Swedenborg. He says that although we cannot deny the possibility of such events, we must investigate carefully, for we know little about the nature of spirits, if indeed there are such. In the past, he was inclined not to accept the credentials of such alleged happenings. However, he expresses a strong interest in the Swedenborg cases, and obviously went to some trouble to verify them as soundly as he could. He also expressed an interest in interrogating Swedenborg himself on these matters. So we know that Kant was fascinated by the reports, sought more information, and reserved judgment, as indeed we would expect from a circumspect reasoner like Kant.[12]

The letter to Miss von Knobloch would lead most of us to think of Kant, the rational scientist, as taking an interest in any fascinating event that requires a difficult explanation. He had already shown that he was not beyond speculation of a marvelously imaginative sort. Thus in his 1755 treatise on the nebular hypothesis, the *Allgemeine naturgeschichte und Theorie des Himmels, oder Versuch von der Verfassung und dem mechanischen Urspringe des ganzen Weltgebäudes nach Newtonischen Grundsätzen abgehandelt*, we find references to "various classes of intelligent beings," "kinds of thinking natures," and to the various "dwelling-places" of these "intelligent creatures." He writes about the dependence of the "spiritual faculties" of the various inhabitants of planets on grosser or finer, heavier or lighter, matter as determined by the "distance of these abodes from the sun." The inhabitants of Jupiter or Saturn belong to the "most exalted class of intelligent creatures. These at least have a different Time-idea from ours; they are not subject to death in the same degree that we are ..." Man occupies a middle ground between these most excellent and the more imperfect grades of "thinking natures."[13] We know that in the middle decades of the century Kant was still struggling with the question of a proper method for metaphysics, and that he had not yet freed himself from the spell of Leibniz. Although Leibniz had insisted that investigation of nature should proceed mechanically, he also believed that one could account for the nature of things according to principles of perfect reason, principles that taught us that reality is a continuity of nonextended simple substances that are internally teleologically oriented and organized. The "double government" methodology of Leibniz was obviously still in Kant's mind during this period, and indeed it never

[12] Not everything Kant says in the letter is reserved. He says of the incident of the Stockholm fire, "The ... occurrence appears to me to have the greatest weight of proof, and to place the assertion respecting Swedenborg's extraordinary gift beyond all possibility of doubt."

[13] I am referring to the neglected third part of this celebrated treatise, *Enthalt eine Vergleichung zwischen den Einwohnern der Gestirne*, where Kant's thinking extends beyond the mechanistic framework of the rest of his book. He claims for these speculations only that they are not arbitrary, though he also adds that the claims are "not quite indubitable." The publisher of this book declared bankruptcy, and nearly all copies were impounded until 1765 or 1766. There is very little evidence that Kant's close contemporaries knew much about his important astronomical speculations.

left his thought throughout his entire career. The speculations in the third part of *Allgemeine Naturgeschichte* are typically Leibnizian.[14]

Kant and the Double Government Methodology of Leibniz. I want to dwell on this Leibnizian background of Kant's thought for a moment, because I believe that Kant never abandoned Leibniz' intent to harmonize teleological and mechanical forms of explanation, and because I think Kant's acceptance of the "double government" methodology provides the clue to an understanding of his apparently changing attitudes to Swedenborg. As we will see, this is a clue that Broad did not pick up in his own attempt to understand these attitudes. Kant's acceptance of this methodology also provides us with access to a prominent feature of his meta-reason, his arational motivation, for concluding that Swedenborg's ideas must be taken very seriously, so seriously as to be worth protecting from frivolous distortion in the public mind.

In many places in his writings Leibniz undertakes to develop and defend what he refers to as the mechanical method. Briefly, this is a ratiocinative and experimental strategy for explaining the behaviour of bodies. Physical phenomena are to be reduced to forms of expression in terms of magnitude, figure and motion. The model of physical interaction is Cartesian; things collide with one another. Leibniz recognizes, however, that since for him bodies are not quite real, this mechanical method will not serve to explain real substances (the monads), which do not collide with one another, and must be understood in terms of another model: that of perception and appetition. this metaphysical method construes substances teleologically, rather than with reference to efficient causes. Mechanical causes are replaced by reasons.[15]

Leibniz gives a rather full statement of what is involved in this methodological dualism in *Critical Remarks Concerning the General Part of*

---

14 It is an historical curiosity that Swedenborg himself had introduced a form of the nebular hypothesis 21 years prior to the appearance of Kant's *Allgemeine Naturgeschichte*, in the *Principia* of his three volume *Opera Philosophica et Mineralia* (1734). Beginning in July, 1736, Swedenborg travelled extensively in Europe. In the notes on his journeys, he records that sometime in 1736 he began the study of Wolff's philosophy, had made his acquaintance, and corresponded with him from time to time. I am not a close student of the Swedish magus, and so I do not know how deeply Swedenborg was influenced by Leibniz, if he was at all. The *Principia* gained for him a European reputation. There is no evidence that Kant had read the work. Kant is not always faithful in naming sources, but in the case of the *Allgemeine Naturgeschichte* he is careful to say than an account in the Hamburg publication, *Freie Urteile* (1751), of the work of Mr. Wright of Durham, had suggested that the fixed stars form a system resembling that of the planets. [The Hamburg account appears in translation in full in *Kant's Cosmogony*, trans. W. Hastie; revised and edited, Willy Ley (New York 1968), pp. 166-79.] I think it fair to assumew that if Kant had known Swedenborg's work he would have credtied it. Finally, it was not Swedenborg's natural philosophy, but his commitment to the paranormal – which he took to be quite normal – that became well-known in northern Europe.

15 I supply some of the details of this summary of Leibniz' empirical methodology in "Leibniz on the Side of the Angels," Chapter II of my *Kant and the Double Government Methodology*.

Descartes' "Principles".[16] Leibniz challenges Descartes' claim to the exclusivity of the mechanical method as the method of adequate explanation, and urges inclusion of teleological factors in metaphysical explanation. Leibniz accepts that the mechanical method is fully adequate to explain the causes of behaviours of material objects, and that physics requires no access to occult qualities or supersensible causes. Nevertheless, causes of material object transactions are themselves derived from "higher" metaphysical causes, to which appeal must be made to sustain complete explanations of the nature of things. This explanatory dualism leads Leibniz to this extraordinary conclusion:

> Nature contains, so to speak, an empire within an empire, or a double government: the government of reason and the government of necessity, or the empire of forms and that of material particles .... Thus, if you study the necessary connection of material phenomena and the order of efficient causes, you will find that nothing happens without a cause that satisfies the imagination and nothing escapes the mathematical laws of mechanism. If, on the other hand, you contemplate the golden chain of ends and the sphere of forms which constitute, as it were, an intelligible world, you will recognize that, due to the perfection of the supreme author, the apexes of ethics and of metaphysics coincide so that nothing happens without supreme reason.

Leibniz' attempt to harmonize both mechanical and teleological forms of explanation by appeal to a rational metaphysics is mirrored in Kant's critical attempt to foster an epistemology and a methodology that also does justice to teleology in the face of a preferred mechanism for explaining the physical world.

Thus the Double Government Methodology (DGM) is also the methodology of Kant, and of Swedenborg (and of most of the rest of us). I need to show in some greater detail the depth of Kant's commitment to DGM. Perhaps it is best to ask; just what parts of the methodology of Leibniz did Kant accept? 1) DGM commits Leibniz to acceptance of mechanical explanation of the behaviour of material objects. Kant accepts this. 2) That form of explanation argues for the elimination of occult qualities and the abandonment of theological appeals in science. Kant accepts this. 3) DGM introduces the duality of reason and natural necessity. The dualism will be greatly transformed by Kant, but he accepts it nonetheless. 4) DGM provides for the separation of science and morality, and identifies the ethical and the metaphysical. Kant's mature works endorse the separation and seek to justify the identification. That his early orientation inclined him to accept this methodology does not surprise us. Kant's Wolffian dependence upon Leibniz is quite obvious in the early works. But we continue to find the impulse to accommodate both mechanism and teleology in the *Träume*, in the *Inaugural Dissertation* of 1770, in the great 1st *Critique*, in the mature ethical treatises, and most clearly articulated in the *Critique of Judgement* of 1790, the work in which Kant finally confronts the threatened antinomial clash between

---

[16] A translation of the relevant section is by Paul and Anne Schrecker, *Monadology and other Philosophical Essays* (Bobbs-Merrill 1965), pp. 78-80.

mechanism and purposiveness, and works through to his form of DGM based on regulative employments of explanatory maxims.[17] For present purposes it is not required that I trace Kant's career-long struggles to ground DGM, a grounding finally achieved within the framework of the critical philosophy. Those who would show that Kant is heavily indebted to Swedenborg normally stress the perceived elements of Swedenborg's thought occurring in *Träume*, the *Inaugural Dissertation*, and the 1st *Critique*. I will have something to say about Kant's use of DGM in these works, but I want to return now to a resetting of the problem in the context of the present examination of Broad's problems with Kant's attitudes toward Swedenborg and supernormal cognition.

From this point on, then, I will take it as given that Kant accepts as an established meta-reason for investigating the problems he does, in the forms in which he does, some strong form of DGM. I will return to further details below.

Dreams of a Spiritseer. I turn then to a discussion of the *Träume*. The work is in two parts. Part I investigates the metaphysics of disembodied spirits and contains both a remarkable theory of matter and a contemtuous examination of the claims of those who profess communication with inhabitants of the spirit world.[18] Part II is in three sections, the first of which deals with the three stories about Swedenborg that Kant had written about in his earlier letter to Miss von Knobloch. The second section discusses the doctrines of Swedenborg that are apparently based on his personal conversations with spirits and his extrasensory perception. Kant's account is full and clear, especially for one based on the *Arcana*. Kant had been disappointed that he received this work, rather than an expected different one dealing directly with theoretical features of Swedenborg's observations. The final section of the book deals briefly with the utility of undertaking serious study of supernormal cognition. In short, Kant suggests that we leave ESP and related paranormality to the dreamers, and get on with the serious business of life.

Despite its playful tone and its unaccepting attitude toward Swedenborg's *Schwärmerei*, the book is a remarkably faithful statement of Leibniz's DGM adapted to a metaphysics in which man is seen as a member of two worlds, one a

---

17 Kant is in a strong sense *the* philosopher of method. I would be prepared to argue that the vital nerve of his lifelong ambition is the quest for a rational accommodation of both Cartesian science and Christian longing for a better life by means of a defensible reformulation of DGM. What I have written about Kant's methodology is summed up in "Kant and the Problem of Scientific Methodology," *Proceedings of the Ottawa Congress on Kant*, ed. Pierre Laberge, Francois Duchesneau, Bryan Morrisey (Ottawa 1976), pp. 222-33. The papers on Kant's teachings on hypotheses cited in the Ottawa paper deal with Kant's attempt methodologically to eliminate a science of the supersensible: "Hypothesis and Explanation in Kant's Philosophy of Science," *Archiv für Geschichte der Philosophie*, 43 2 (1961); "Kant on Hypotheses in the 'Doctrine of Method' and the Logik," *Archive für Geschichte der Philosophie*, 44 2 (1962). These papers are now supplemented and improved by the full treatment of methodology in my *Kant and the Double Government Methodology*.

18 Broad's analytic account of the theoretical parts of the *Träume* is extensive and quite excellent; I cannot improve upon it here.

world of body, the other, of things of the mind, of which ethical pursuits are for Kant here, as elsewhere and subsequently, uppermost. He is careful to point out that the problem of the existence of spirits – the metaphysical problem – is recalcitrant to direct management by means of DGM. This is because everyone seems normally doubtful about the existence of spirits; thus the existence of spirits is not suggested by experience. Rather, "... many conceptions arise in secret and obscure conclusions incidental to experience, and afterwards are transmitted to other minds without even the consciousness of that experience or conclusion which has first established the conception. Such conceptions may be called 'surreptitious.'" This concept of a surreptitious cognition is the prototype of what Kant will call a 'subreptic axiom' in the *Inaugural Dissertation*, an axiom derived by misusing a sensitive concept as an 'intellectual phenomenon.' Of course for Kant the latter phrase is a contradiction in terms.[19]

Apart from its nonstandard and unscholarly style, and the fact that it was published anonymously, what is there about *Träume* that makes many question why Kant troubled to write it? It seems to deal with especially troublesome and "unscientific" matters. It is suggested that in Kant's time and place taking paranormal phenomena seriously is equivalent to experimenting with LSD at Harvard.[20] Kant was not yet a professor in the Albertus University of Königsberg, but only a *privat dozent*. Thus, quite apart from the fact that the little book contains a compact form of views that Kant will later give expression to in "scientific" works, it seems reasonable to wonder why the untenured teacher would risk all in writing it.

We thus have two questions to confront: why did Kant write the book at all? and why did he publish it anonymously? My answer to the first is that it is a

---

[19] The terms 'subreption' and 'surreption' have a common root, and both have the sense of false, sneaky, deceitful. All of this is likewise true of Kant's German term, "*erschleichen*". An interesting early meaning of 'surreption' has to do with a sin or a temptation secretly stealing unbidden and unexpected into one's mind. our words also share a common root (meaning 'to creep') with 'reptile' and 'serpent'. One should bear in mind in all of this that the *Critique of Pure Reason* abounds with discussion of what happens when we seek to extend cognition beyond the limits of experience, when, so to speak, idle and inutile thoughts 'creep in'. I think it likely that Kant took this quasi-religious sense of bad epistemology quite seriously. To mark and dispel surreptitious cognition became for him, as we shall see, a kind of methodological crusade.

[20] I have mentioned Kant's later interest in the *Schwärmerei*, the irrational enthusiasm for things paranormal that seems to sweep through northern Europe in the second half of the 18th century. Hamann, Kant's sometime friend and proofreader for the 1st *Critique*, writes in a letter to Kriegsrath Scheffner that he regards Swedenborg's spiritual paroxysms as a form of "transcendental epilepsy" producing critical frothing (Nov. 10, 1784) There seems to have been a readiness on the part of many to regard the *Schwärmerei* as a form of mental illness. For Kant the *Schwärmerei* is a corrupt mental state in at least three senses: it is a sickness, a distorted form of cognition, and an inappropriate ethical and religious enthusiasm. Hamann adds in his letter that the entire theory of evidence of the church is derived from this transcendental epilepsy. He concludes, "Unfortunately dreams and sicknesses are the best data of the energy of our souls." We will see below that this problem of data is for Kant a central concern, a concern fully expressed in an important letter to Moses Mendelssohn.

book on metaphysics embodying Kant's continuing interest in locating the methodological limits of empirical knowledge. Given *that* context, Swedenborg's claims to ESP, precisely because they were made by a well-connected scientist who was himself of widespread repute, were exactly the kinds of claims to take seriously as test cases of Kant's sophisticated recasting of the double government methodology. My answer to the second question is that although the book appeared anonymously, Kant took special pains to get it into the hands of just those official and semiofficial persons who could have done him the most harm, and not to try to hide authorship of the book from them. These answers lead me to propose a modified Gutting Type I explanation for Kant's interest in the paranormal. But first, some consideration of Broad, who proposes a Gutting Type II explanation in precisely the kinds of terms that are grist for the mill of advocates of the strong programme.

C. D. Broad's Sociological Explanation. Instead of a volume by Swedenborg presenting and arguing for his claims to seership, Kant received the eight volumes of the *Arcana Coelestia*. Having waded through these ravings of a "transcendental epileptic," a bored and offended Kant struck back. Now Broad says that this explanation is offered "very tentatively." Broad knew what is in the book, an examination of the limits of empirical cognition. Nevertheless, his suggested psychological explanation, though very tentative, must be taken seriously, because we are trying to account for adoption of a literary style foreign to Kant. Kant wanted to prevent as many as possible from taking too seriously the substantive claims of Swedenborg, who Kant thought had mislocated and misdescribed the intellectual world.

It is Broad's extraordinary sociological explanation of the need for anonymity that mainly distresses me. This explanation is as follows: "It is reasonable to suspect that, in 'enlightened' academic circles in East Prussia in the middle of the eighteenth century, a reputation for having carefully read Swedenborg's writings and having paid serious attention to the evidence for his alleged feats of clairvoyance, would be enough to condemn a *privat dozent* to remain in that position for the rest of his life.... I can only suggest that the conventions of the time and place permitted a *privat dozent* to flirt with this disreputable subject, provided that he made an honest man of himself by maintaining the form of anonymity and by adopting a sufficiently bantering and condescending tone towards the alleged phenomena and the persons of whom they were narrated. If these were the conditions, Kant certainly complied with them."[21] Now that is as good a Gutting Type II explanation as one could wish. It is, however, false.

The case for Broad's sociological explanation is considerably weakened by imperfections in his intellectual history. He thought that Kant's references to Swedenborg were limited to the letter to Charlotte von Knobloch and the *Träume*.

---

[21] Broad (1953), p. 127.

but there are other materials to which I have referred, including the two letters to Mendelssohn, and for reasons I will now detail these must be taken very seriously. In the letter of february 7, 1766, Kant says that he is sending Mendelssohn copies of "some 'Träumereien'", one for Mendelssohn himself, and for six others. He says of the work: "It is a hastily prepared treatise, and presents rather an outline of the manner in which questions of this kind may be treated than the treatment itself."

Kant repeatedly sought Mendelssohn's approbation of his works in metaphysics; there can be no doubt that he sought it in this case. What of the six others? Two are Court servants, Court Preacher Sack, and Councillor of the Consistory Spaulding. I have made no effort to trace these gentlemen, but it must be noted that Kant made no effort to hide his impious book from two persons obviously highly placed in religious officialdom. If anyone could threaten his tenure in the university, these two could. the other four persons are all members of the Berlin Academy and hence also royal appointees (Mendelssohn had been proposed for membership, only to be blackballed by the King himself). These gentlemen are Provost Süssmilch, and Professors Lambert, Sulzer, and Formey. Johann Heinrich Lambert was regarded by Kant as "the leading genius of Germany." He was the only member to be admitted to all four classes of the Berlin Academy: mathematics, natural philosophy, speculative philosophy, and letters. Kant had the highest regard for his work, and for his opinions. Johann Peter Süssmilch gained some repute for defense and development of Hamann's theory that nature is *logos*, believing that humans learned language by "divine instruction." He was the target of Herder's critique of the divine origin of language. Both Kant and Swedenborg thought of man's response to the supersensible as *symbolical*, as in some sense linguistic. Perhaps Kant thought that his little book made a modest contribution to the disputes about the origins of language.

Samuel Formey was the permanent secretary of the Berlin Academy, and a popular interpreter of the philosophy of Wolff. He contributed to the widespread feminine interest in Wolff's thought with his 6 volume *La belle Wolffian* (1741-53). The author and editor of nearly 600 books, he sought to cancel the boredom of his declining years by issuing his own *Oeuvres posthumes*! As permanent secretary of the Academy he wielded considerable influence, and was thus a figure Kant needed to take seriously. Johann Georg Sulzer, a Swiss, was a very prominent member of the Academy. With Euler, he had sponsored Lambert for membership in the Academy. In his own philosophy he sought to replace ideation with feeling in working out the development of the Wolffian philosophy, a move which probably impressed Kant favourably. He translated Hume into German. He created the essay competition of 1763, won by Mendelssohn, but resulting in the publication of Kant's second-place effort now known as the *Prize Essay*. Kant often sought his response to his work, a response that was frequently quite limp, as in the case of his reaction to the *Inaugural Dissertation*.

This must count as an impressive array of highly-placed individuals to whom to send copies of an anonymously authored book. If Kant had really intended to keep a secret of his authorship of the *Träume*, he did nothing that I could learn to promote that secret. Instead, he seems actively to have sought reactions from prominent churchmen, academic politicians, and quite important philosophical scholars. Thus, however one decides the question of anonymity, I think we must conclude that Kant took the message of his book very seriously and wanted that book to be regarded as an important contribution to the studies he was making into the foundations of metaphysics. Further confirmation of this thesis is obtained from Kant's second letter to Mendelssohn of April 8, 1766, to which I now turn our attention.

The Second Letter to Mendelssohn. Kant begins by thanking Mendelssohn for forwarding the copies of the book. He next expresses regret that Mendelssohn was offended by the tone of the book, a tone that Mendelssohn found "between jest and earnest." I will discuss the rest of the letter in some detail, because in these comments to Mendelssohn Kant decisively makes my point that he was motivated to write the *Träume* by a cluster of meta-reasons stemming from his search for a correct form of the Double Government Methodology. If I am correct in this, Broad's Gutting Type II sociological explanation is not only wrong, but an impediment in the way of a correct understanding of Kant's interest in the paranormal.[22]

Kant's second letter to Mendelssohn seems to me to answer all of the remaining questions surrounding his book on Swedenborg. Clearly, he wrote the book with some reluctance, and in a state of confusion not characteristic of Kant. the actual production of the book seems to have been astonishingly casual, if we believe Kant that it was printed a proofpage at a time, so that one sheet would be off at the printer's while Kant worked on the next. Nevertheless, reluctant, confused, and casual as Kant's attitudes and actions may have been, it seems very clear that he thought the object of the exercise worthwhile, if not imperative. That object was to call into question both a method – the "dream science" – and the alleged cases of paranormal cognition. It is the "contagious," hence irrational, unhealthy, nature of the *Schwärmerei* that so distressed Kant, especially at a time when his own sceptical yet positive critique of metaphysics was in the early stages of development. In short, he took both Swedenborg's ponderous science of the supersensible and his reputation as a storied clairvoyant to be threats to the Double Government Methodology as a scientific manifesto and as ultimate protector of public morality. Kant's interests as revealed in the *Träume* are soteriological in a suitably secular sense.

Kant says in the letter that he is seeking a "specific criterion," presumably one that will systematically rule out cases of supposed direct access to spirits. That criterion, and its attendant doctrine of method, will appear in mature dress in several forms in the 1st *Critique*. In general, it stems from Kant's insistence on separating questions of knowing (within the bounds of possible experience) from questions of pretended insight through intellectual intuition (surreptitious cognizing once again). The specific rational claims to direct insight into the soul are of course

---

[22] The letter is translated in Arnulf Zweig (1967), pp. 54-7.

examined by Kant at length in the paralogisms section of the *Critique*, where the surreptitious pretense of rational psychology is fully exposed. Kant's methodological conclusions will be harsh: in the 1st and 3rd *Critiques* he will insist that the existence of spirits cannot be hypothesized, cannot even be held to be likely or probable (*or* unlikely or improbable), because spirits are *categorial impossibilities*, not knowable within the bounds of possible experience.[23]

Kant tells us in the *Träume* that the origin of the belief in spirits is surreptitious. We learn more on this topic from the letter to Mendelssohn. Here Kant says directly that the central problem is one of data or evidence that would give assurance of the way in which the soul is present in the world, and, most crucially, he takes this problem (of the union of soul and body) to be a sub-case of the general problem of any kinds of connection between external powers and faculties that are merely receptive. This, of course, is *the* problem of the *Critique of Pure Reason* – we may fairly take it to be the single most important question Kant faced in his long career. In the letter he also mentions what will become the motivating theme of the Dialectic of Pure Reason: can inferences of reason (he will later add, "unaided by intuitions") discover "primitive powers," that is, real causal powers? The pre-critical *critical* note is sounded in the final quoted sentence of the letter: "Here we must decide whether there really are not limitations established by the bounds of our reason, or rather, the bounds of the experience that contains the *data* of our reason."

The question of data, the question of evidence – the question that is the nerve of DGM: the divide that finally cannot be crossed, the limit of the knowable. The philosophical crucifix. Kant was putting Swedenborg to the philosophical experiment of the cross. That, as I read the evidence, is the message of the *Träume*. It is a message fully consistent with Kant's philosophical worries *throughout* his career. It is a message revealing the enthymematic premise that functions as the meta-reason for Kant's troubled attitudes toward Swedenborg and the paranormal. That Kant's style in dealing with spiritualism is partly playful and partly scornful seems to me to present no special problems. It is not easy for a serious philosopher to drive the nails into the hands of a distinguished and popular foreign colleague. The more telling point, however, is that the question of Kant's atypical style is seriously diminished in importance once we understand how central a role DGM plays in Kant's *Träume*.

As I understand these matters, the two letters to Mendelssohn that he apparently did not know about, coupled with the direct availability of employment of the Double Government Methodology in *Träume*, provide a decisive refutation of Broad's sociological/psychological account of Kant's interest in the paranormal. I also take it that I have made part of Laudan's point: the intellectual history, properly worked out, cancels the need for sociological intrusions into explanations of scientists' beliefs, in just those cases where arational factors are intermixed with,

---

[23] See the references to my studies of Kant on hypothetical reasoning cited in footnote 17 above.

or presupposed by, the scientists' accepted forms of reasoning. The arational factors, thus understood, can be ingredients in a rational reconstruction of scientific acceptance. I take it to be the central task of the historian of ideas to locate such arational factors – as I have tried to do in the present study of Kant – wherever possible, and to give guidelines concerning the incorporation of these factors into a rational account of scientific belief. This is not to deny that sometimes (often?) arational – even irrational – factors play a role in the beliefs of scientists; it would be strange if this were not so. Nor is it to deny that sociological explanations of science as institutionalized behaviour is impossible. It is just to deny the primacy of sociological explanation at every level. If my examination of the case of Kant can be generalized to other cases, as I believe it can, then internalist explanations of belief can often take the form of appeals to arational factors as meta-reasons within the larger context of reasoned cases of acceptance. Such factors tell us important things about what a given scientist takes to be at stake in pursuing the line that he does. And where, as in the present case of Kant, the utilities involved in a single piece of work find more extensive use and expression in other parts of the scientist's endeavors, the case for inclusion of arational commitments as meta-reasons in the reasoning becomes compelling.

Träume and the Inaugural Dissertation. To complete my study of the case of Kant and ESP, I have, then, one more task to perform. I must show, at least in outline, that indeed Kant's *Träume* commitment to DGM did continue to motivate his further researches. I have mentioned some connections between the problem of spirits that preoccupied Kant in *Träume* and his treatment of that problem in the *Critiques*. It will help to make my case for Kant's abiding commitment to DGM if I say a few words about continuation of his interests in the pre-critical period, interests revealed fully in the *Inaugural Dissertation*, and a few words about his positive development of interest in spirits given expression in his introduction to the idea of a *corpus mysticum* in the 1st *Critique*.

It has not been my aim in this study to expose the details of Kant's *Träume*, but I want now to say something more about his theoretical and especially methodological conclusions in this work. After showing that the presumed power of spirits to act on one another and on bodies cannot, like physical force, be located, Kant next develops a long story about what it would be like to be human and a member of two worlds, one physical and the other spiritual. Such a human "ghost-seer" would possess all manner of paranormal possibilities of cognition, would, in short, have ESP. The "communication" between seers could not be direct, because there is no intuitive sensory content in paranormal cognition; it would therefore be *symbolic*, largely a matter of interpreting signs given either as material objects or as words. The communion of spirits could not be located or intuitively represented except in quasi-sensory ways. Intelligibility in the world of spirits would be a matter of gnostic privilege in reading symbolic messages. A person like Swedenborg, who takes himself to be a member of such a spirit world (while at the same time being a member of the world of bodies, is either a genuine spirit-seer, or is in a pathological state of delirium. Kant suggests that the two states are really indistinguishable, and that one who chooses to think of all alleged cases of ESP as pathological has certain advantages, for example, he can easily dismiss all allegations of the paranormal just because they fail to comply with ordinary demands of coherence in common experience.

Kant's theoretical conclusions are two: [1] the paranormal and the physiological pathological are not clearly distinguishable; and [2] we are, and are doomed to remain, wholly ignorant of solutions to three basic problems: (i) the problem of the animation of a human body by a human soul at conception; (ii) that of the continuing connection between a soul and a body during life; and (iii) the separation of the soul from the body at death and its possible continuing post-death existence. His methodological conclusion points the way to further developments in the critical philosophy. To be genuine, a scientific hypothesis must deal with possible objects, and must speculate about detailed features of the behaviour of such objects. However, we can only know about possibility given actuality, and the only actual objects we have available to us are encountered in sensation. Possibility must have intuitive content (hence the need for Kant later to appeal to constructions in mathematics which give intuitive content). His methodological conclusion in *Träume* is therefore sceptical and agnostic, as he told Mendelssohn in the second letter it had had to be. Both the theoretical and methodological conclusions of this work are consistent with Kant's adoption of DGM. Science is limited in its investigations to the physical world where the connections between objects are mechanical and deterministic; the other "world," the world of cultural presence, of morality, art, and law, is a world that can only be understood by perfect reason – which for Kant will finally mean *practical* reason.

I turn now to Kant's *Inaugural Dissertation*, defended in 1770, just four years after the appearance of *Träume*. I take it that the main themes of the dissertation were in Kant's mind when he wrote the second letter to Mendelssohn, that is, in the year of the appearance of *Träume*. It is of course now well established that the dissertation is the first of Kant's "critical" works, containing important new ideas the full working out of which would occupy Kant for the next decade. What is not well recognized is that the main lines of argument in the dissertation follow more or less exactly upon the conclusion of *Träume*, and amount to a first attempt on Kant's part to solve the general problem of any form of external impact on receptive faculties. He had told Mendelssohn, moreove, that the central question is one of data; in the dissertation, we find Kant's first appeal to the singularity and immediacy of data or evidence in the introduction of this special concept of an "intuition".

The *Inaugural Dissertation* begins by an investigation of the properties of "a world."[24] No one could accuse a young follower of Wolff of idle curiosity in further pursuit of understanding which worlds are possible, which actual. But

---

[24] My references to Kant's *On the Form and Principles of the Sensible and Intelligible World* are to the translation of G. B. Kerford and D. E. Walford, *Kant, Selected Pre-critical Writings* (New York 1968).

Kant's interests turn out to be rather special; his thought in the dissertation already takes the epistemological turn – his deep question concerns which worlds are knowable. Kant begins the discussion in a typically scholastic way (remember, this work is one Kant had to defend publicly and show to be fully compatible with current modes of doing philosophy).[25]

Briefly reviewed, this is what the exercise of Kant reveals: *Analysis* arrives at a part which is not a whole, a *simple*. *Synthesis* arrives at a whole which is not a part, a *world*. The logically most primitive idea of a world (call it World I), is that of a composite, or set: from a numerically distinct collection of parts, the intellect forms the abstract notion of whole. There may be many such worlds; being creatures of the intellect, they are governed solely the law of non-contradiction. Kant takes the problem to be one of representing this abstract notion by means of a distinct *intuition*. Call this intuitively presented world World II. Such a world, Kant thinks, would be subject to certain temporal conditions, as well as logical ones, the temporal conditions giving grounds for *successive addition of parts to generate a whole*. Being subjected to time means being subjected to intuited singularities coming one after another; singulars, not wholes, are given. His question now becomes: how many worlds of World II type can there be? To answer this question we must look at two of the special kinds of properties he thought of as characterizing worlds.

The *matter* of a world is its parts, and these parts are *substances*. That which endures through change is a part of a world. Interestingly, then, a world is not itself a substance, but a composite of substances. No substance is for Kant a world. For example, he says, "the so-called world *of the ego*, which is completely constituted by a unique simple substance together with its accidents is not properly called a world, unless it is called an *imaginary* world."

The *form* of a world is a bond established by the principles of "the *possible influxes* of the substances which constitute a world." Actual influxes are *states* of the world, and "transeunt" forces, the causes of the influxes, presuppose a principle by means of which each independent substance is related to others. In the absence of such a principle, the positing of transeunt forces would be impossible. A world form is thus a set of possibilities maintained throughout changes in the actual states of the world. The identity of the parts is not sufficient for the identity of the whole; the identity of the whole is only insured by the principle of composition. Each "real" world will be given in a form that is constant and invariable, and this form will act as the "perennial principle" of each contingent and transitory form that is a *state* of the world. Kant concludes that the actual world is the world whose changing states are invariant with respect to the forms of space and time, as precisely the forms of receptivity of all singulars of intuition. The compositional identity of World II is space/time identity. It must be

---

[25] I am not proposing a sociological reason for Kant's philosophical style in this work, apart from the explanatorily neutral observation that when engaged in a ceremony one usually follows the rules, or else is not taken to be engaged in that ceremony.

noted, however, that the states of possibility are only generated by *given* forces, actual forces empirically ascertained. Kant's insistence that forces can only be given in experience, and cannot be cognized *a priori*, is urged in all stages of the development of his philosophy. This being so, can one really wonder why he had to take seriously the claims of a major European scientist to direct access to the spiritual lives of others?

In 1770 Kant's theory of worlds yields the following structure:

1. Logically considered, the members of a world consist of a set of self-dependent substances.

2. A logical set of substances is not itself a world. A world is a composed or related set of substances, and the relations binding substances together in a world can only be external relations.

3. The form of such external relatedness is a set of possible interactions (influxes).

4. In any world accessible to human knowers the actual interactions between substances are empirically noted, and are *states* of the world. In the *Dissertation* the distinction between substance and state, between form of world and principle of determining state of a world marks the new departure from the Leibnizian concept of world in *Living Forces*.

5. Whatever transeunt forces we postulate as causes of states of the world must comply with the form of the world.

6. The form of a world is not given independently of the world, but is a condition on what it is possible for us to know in and of that world.

7. The interdependence of self-subsistent substances is guaranteed as a causal nexus supplied by the forms of space and time.

8. These forms do not determine the content of a knowable world, but insure that any forces to be countenanced in the sensible world must be given in empirical intuition.

9. The form of any knowable world is space/time. The principle associated with this form rules out intellectual intuitions and spiritual forces, except as logical possibilities.

10. World I is thus uninstantiated. Its form is that of deductive logic. It exists only as a logical possibility.

How many worlds are there of type II? Only those for which there are forces. Kant had written in the second letter to Mendelssohn that his general problem was one of understanding any case of external impact upon that which is merely receptive. External impact is exerted through the transeunt forces,[26] and these can only be known through sensation organized by the receptor forms of space and time. Kant's World I, as a creature of pure thought, is symbolically cognized (sect. 10); it cannot be represented intuitively – we have no data to instance it. The matter of our cognitions is only given in sensation. World II is thus the only world that can be actual *for us*, actual as a *phenomenal* world or sense-intuited objects organized in space and time, and driven by physical forces empirically ascertained. Kant's important conclusion in the dissertation regarding evidence or data – the central problem marked in the letter to Mendelssohn – is that there are two acceptable forms of intuition. Immediate singulars are given in *sensory intuition*, and in *a priori* or *pure intuition* (a form of direct presentation of particulars later to be rendered by Kant as mathematical *construction*). There is no *intellectual intuition* that would give us immediate access to World I.

The direct bearing of Kant's first attempt at a solution of his "general problem" on the question of the existence of spirits (and hence the knowability of spirits) is not far to seek. The presumed existence of spirits is introduced surreptitiously, by inappropriate – not intuitively instantiated – appeals to rational principles, for example, the principle of non-contradiction or identity on which Leibniz had placed such great emphasis. In *Inaugural Dissertation* (sect. 28) this is what Kant says:

> But by treating the subjective conditions of judging as objective it is rashly concluded that for that reason it would not be permitted to any intellect at all and so, that *whatever does not involve a contradiction would be therefore possible*. Hence so many vain fabrications of *forces* I know not what invented at pleasure, which freed from the obstacle of inconsistency burst forth in a horde from any spirit which is architectonic or if you prefer it with a proclivity for chimeras. For since a *force* is nothing else but the *relation* of a substance A to *something else* B (an accident) as of the ground to the grounded the possibility of each force *does not rest upon the identity* of cause and caused or of substance and accident. And so also the impossibility of falsely fabricated forces *does not depend upon*

---

[26] It is important to translate Kant's Latin "transeunt" with the same English word, and not to use "transient". The Insel-Verlag German translation gives "übergehende", which is literal and helpful. The OED provides justification for my suggestion, pointing out that "transeunt" has the force of the opposite of "immanent". The transeunt forces operate beyond themselves, causing the influxes. Kant's point is exactly that such transcendent or transeunt forces can only be posited to account for that which is spatio-temporally ordered. The "actual influxes" are thus sensuously ascertained terminations of the work of transeunt forces, and the "possible influxes" will later become Kant's "objects of possible experience". Although he is careful to limit positing of these forces within space/time formed world II – thus eliminating the specious postulation of objects of intellectual intuition and an inhabited spirit world – Kant does not remove the suspicion that he is countenancing the existence of objects that are causally efficacious within experience, and yet which actually transcend experience, until the 1787 2nd edition version of the *Critique of Pure Reason*. Nevertheless, Kant held throughout his career that the only forces we can know are forces given empirically (in *Living Forces* they are known by their effects); see, for example, A207/B252 in the 1st *Critique*. Kant's treatment of worlds in the Dissertation is part of his attempt in the 1760s to bring forces within the limits of sense experience, and under the constraints of the epistemic grammar of space and time.

> *contradiction alone*. So it is not permitted to take any *original force* as possible unless *it has been given by experience*, nor can its possibility be conceived *a priori* by any perspicacity of the intellect.

Kant's ultimate appeal to well-ordered judgement under the strictures of DGM occurs in sect. 30:

> ... the ill-considered citation of things supernatural is the couch upon which reclines a lazy intellect. For the same reason, *comparative miracles*, namely influxes of spirits, we carefully keep out of the exposition of phenomena. For since their nature is to us unknown the intellect would be turned aside to its own great detriment, away from the light of experience by which it alone has the opportunity of procuring for itself the laws of judging, towards shadows of species and causes to us unknown.

The only worlds we can know are worlds forced upon us. The principle of non-contradiction (identity) can only establish logical possibility, not real possibility. Leibniz knew this, and sought to have compossibility supplement logical possibility. For Kant, the primary criterion of compossibility is the order of actual influxes. We have no experience of spiritual influx, therefore there is nothing to ground a spiritual world except as a logical possibility, a possibility of pure reason. For Kant, the intellectual world (World I) stands only as a kind of logical surrogate for a world of actual spirits. The cognitive status of that world must therefore be radically rethought, a task Kant undertakes in the *Critiques*.

Note the interesting twist. By an application of his special concept of a possible world Kant effectively eliminates the spirit world as a possibly actual one. Possibility can only be established by actuality (perceived forces), and this is why Kant insists that there be some data to underwrite the claims to the existence of a spirit world – this I take it is why he was so fascinated by the alleged clairvoyance of Swedenborg; he wanted to learn if these cases are authentic, because if they are, he would have to give up his moves away from Leibniz, and conclude that two structures of possibilities – two worlds – are actualized, the phenomenal and the noumenal: in other words, he would have to accept the metaphysics of Leibniz on the evidence of direct contact between embodied souls and disembodied ones. What he does conclude – in the absence of evidence of spiritual forces – is that the two worlds follow from two uses of reason, one theoretical, one practical. Perhaps he fails, but the attempt is a consistent application of DGM, and is entirely faithful to the methodological motivations of Leibniz.

The Corpus Mysticum. In the *Critique of Pure Reason* Kant introduces his special *moral world* (call it World III) as a transformed World I and World II combined, or as a form of World I seen from another point of view. The datum of World III is human freedom – human freedom is, if you will, the influx force of the moral world, that which gives it its form. If we subtract from this world all actual moral content, it is an intelligible world only (a world of type I). Kant says:

> To this extent, therefore, it is a mere idea, though at the same time a practical idea, which really can have, as it also ought to have, an influence upon the sensible world, to bring that world, so far as may be possible, into conformity with the idea. The idea of a moral world has, therefore, an objective reality, not as referring to an object of an intelligible intuition (we are quite unable to think any such object), but as referring to the sensible world, viewed, however, as being an object of pure reason in its practical employment, that is, as a *corpus mysticum* of the rational beings in it, so far as the free will of each being is, under moral laws, in complete systematic unity with itself and with the freedom of every other. (A808/B836)[27]

We are now in a good position to see how Kant works out the status of his World III in ways generally faithful to DGM. The interests in that world are typically moral interests; they are subject to the demands of practical, not theoretical, reason. As an idea, the moral world is also an *ideal*: activities in the sensible world are to be adjusted to its demands. We need not stumble over the details of Kant's exegesis of World III in the 2nd *Critique* and the later ethical writings. My point here is just the one that in his attempt to deal with Swedenborg's alleged experiences of the spirit world, and with his spiritual doctrines more generally, Kant was extending his deep commitment to DGM to the fullest measure, even to the point of reaching what Leibniz referred to as the apex of ethics and metaphysics. At one crucial level of his thought, Kant identified the interests of metaphysics with those of ethics. Two appeals to the arational are made in all of this: an appeal to the datum of human freedom (corresponding to the appeal to sense intuitions in the case of World II); and an appeal to the fundamental and unchallenged rightness of the Double Government Methodology.

The Reasonableness of Kant's Interest in Spiritualism. I now take it that my case has been made. A proper understanding of Kant's interest in the paranormal can be had by close attention to certain details of his intellectual career in the 1760s, understood against further developments of that career in Kant's mature critical period. Given this understanding, no appeal to external psychological and sociological factors is required. Kant's final verdict on the Swedenborg allegations *was* equivocal, but in entirely defensible ways, if we take the arational factors into account. There is a World III, but its inhabitants are all living in World II. To take World III to be itself natively inhabited is to take World III to be a World I with intuitive content. For Kant, there is no such World I. However provocatively *Träume* may have been styled by Kant, the message of this book, and its intended impact, is systematically and practically related to the contents of the second letter to Mendelssohn, to the fragments referring to Swedenborg, and to the lecture notes, to the *Inaugural Dissertation*, and to the later critical works. It is clear, furthermore, that Kant *had to be* concerned with claims to ESP. For if there are spiritual forces, if spirits can be given a habitation, then Kant's entire critical programme collapses in shambles.

---

[27] The translation is that of Norman Kemp Smith (London 1950), pp. 637-38.

Replacing the Broad Explanation. I will try to draw together the threads of this complicated challenge to the strong programme in sociology of knowledge. In seeking such culminated interlacements philosophers often appeal to formalisms. In the present case, no exotic notation is required, for we have the Gutting explanatory schemata ready at hand. What needs to be accounted for is Kant's complex attitude toward the question of the existence and knowability of spirits, an attitude seriously coloured by the tone and anonymity of *Träume*.[28] Kant's conclusion may be stated as follows:

1. The existence of spirits is provably unknowable, although that existence is a *logical* possibility. However, claims to the factual existence of spirits must be taken seriously, because such claims are based on a surreption of actual perception, and this procedure must be exposed, isolated, and the damage done by it corrected. [I take it that this is Kant's full conclusion of this matter. The tone and style of the *Träume* appear to call this view into question, or at least give one grounds for thinking that Kant's views on ESP and related paranormal phenomena are equivocal or unsettled.]

A Gutting type II explanation of this conclusion (call it a "Broad Explanation"), offers the following explanans:

1.1 An East Prussian *privat dozent* living in the mid-18th century needed to address himself only to serious and accepted scientific concerns, and needed suitably to disguise any interests in academically marginal or proscribed topics like ESP and the existence of spirits. Failure to comply with such fraternal requirements might well mean non-appointment to a professorship.

1.2 Serious scholars who spend large sums of money for boring books which are not the ones they expected to receive often take a negative view of the sentiments of the authors of such books, and write abusive responses.

---

[28] I think I have sufficiently dealt with the question of anonymity above. If there is a residue to be explained, then *this* is a place for addition of sociological considerations; but such considerations will have no bearing on Kant's beliefs regarding the paranormal. Some relevant features of this adjusted accpunt are the following. In his letter to Mendelssohn of April 8, 1766, Kant says that he wrote *Träume* to prevent people from pestering him with questions about his "prying inquiry into Swedenborg's 'visions'". I take this to be the *occasion* for writing the book, not the *reason*. I have shown that quite a few of Kant's contemporaries knew that he had authored the book. It is one thing to accept authorship of a work dealing with matters of current debate; it is quite another to want one's name fixed in perpetuity to a hastily written book, one with themes not fully developed, one written in an unscholarly style, one in which one's later best arguments are not put forward. These matters receive further treatment in my *Kant and the Double Government Methodology*.

1.3 Kant found himself in both sets of circumstances

I have argued that the Broad Explanation fails to do justice to Kant's attitudes in the *Träume*, and his views on spiritualism in general. The Broad Explanation fails because it is based on an imcomplete historical study, and because it ignores Kant's standing commitment (an at least partly arational commitment) to the Double Government Methodology and its implications. I would therefore propose the following explanans as fully capturing what I take to be Kant's conclusion:

1.4 Kant's explanation of DGM led him to expect that explanation of alleged paranormal phenomena would have to be cast in mechanical terms, or such phenomena, failing of explanation in those terms, would turn out to be products of surreption, meaning that the true status of the phenomena would therefore have to be radically rethought.

1.5 Kant discovered that spiritual phenomena are surreptitious – that the data of the existence of spirits could not be regarded as reliable. Nevertheless, spiritualism has much popular support, and threatens to subvert orderly morality by encouraging irrational motivations.

From which 1. follows.

Why did Kant take the allegations of Swedenborg's spiritualism seriously in the first place? I have dealt with some details of the answer above. The answer involves recognition of the following: [1] Kant's meta-reason, his acceptance of DGM, already committed him to some form of explanation of "spiritual" phenomena; [2] Kant's persistent empiricism led hed him to insist on clear evidences (data) of existence; *because* [3] his motivating problem in the 1760s was one of impact (influx) on that which is merely receptive; [4] the preformation of doctrines later to be given full expression in the 1st *Critique* required a principle of demarcation separating authentic cognition from metaphysical pretense (recall that in the letter to Mendelssohn Kant refers to his new "important insights in this discipline [metaphysics] ..., insights that will establish the proper procedure for metaphysics. My notions are not merely general ones but provide a specific criterion."; and [5] even as late as the 1790 letter to Borowski Kant was disturbed by the *Schwärmerei*, the irrational enthusiasm that threatened to defeat religion within the bounds of reason alone, religion sharing the apex with ethics and metaphysics. We thus have reasons and meta-reasons aplenty to account for Kant's quite legitimate interest in the unseemly side of metaphysics.

Assimilating Meta-reasons into Rational Explanation Schemes. A single case like the one I have just investigated in detail cannot, of course, sustain the general case. Was Kant's approach to the problems of spiritualism "rational"? Yes – and no. It cetainly was rational, and progressive, in Laudan's senses of these terms. It tended, within Kant's enlarged research tradition (including importantly Leibniz' DGM), toward solution of a number of problems that would have remained unsolved within the limits of the alternative tradition (Swedenborg's spiritualism).[29] Still, it must be admitted that some of the features of Kant's approach to these problems were not "internally" rational in the sense that for all of his beliefs Kant had given, or could give, or was prepared to give, a reasoned defense. Some of those beliefs were "basic", or arational. But they go along with the entire programme or tradition, helping to make it serve its larger purpose. Finally, what emerge as constitutive elements in Kant's developing beliefs about the paranormal call for no independent (sociological) explanation, once the intellectual history is written in full detail, as I have endeavored to do. In the best sense of the term I can think of, Kant had *reasons* for taking on Swedenborg, and for developing the view of spirits that he did. None of those reasons were uniquely sociological.

Earlier in this essay I suggested that meta-reasons are utilities, ends in view. In my analysis of the episode in Kant's intellectual life, many such utilities appear. All five of the requirements of Kant's programme outlined above involve utilities. Some are local, like attempted satisfaction of the demands of the problems currently exercising his mind. Some are writ large, like his long-term commitment to satisfying the (religious, ethical) demands of practical reason. On my view of what is required of intellectual history, these utilities become part of the story to be told, *within* the processes of thought of scientists. Finally, I think that Laudan (and James Brown, and others) is largely right in insisting that explanations of scientific belief must conform to some (correct) theory or rationality. If this is *my* methodological meta-reason, so be it.[30] Acceptance of such a meta-reason contributes importantly to awakening from the present *Träumerei* promoted by too much attention to the claims of the strong programme in sociology of knowledge. One would be wrong, however, to suppose that all enthusiasms are pathological. the strong programme is not pseudo-science. In shifting attention away from utilities and meta-reasons to scientifically unfounded generalizations, it presents a methodology that distorts intellectual history.[31] Sociology is by no means redundant

---

[29] Kant's rejection of the knowability of spirits, his deliberate failure to distinguish pathological maladjustment from paranormal experience, and his rejection of any rational psychological solution to the problems of soul/body connection, all pave the way for new departures in both psychology and physiology, not to speak of the liberating effects of his thought on ethical theory and philosophy of religion. Swedenborg's spiritualism generated a church.

[30] James Brown, in the doctoral dissertation cited earlier, seeks to show that such meta-reasons can be rendered as reasons. I am not sure that he is right, but that is a matter for another paper.

[31] I am aware that the case study I have offered leaves some conceptual considerations unresolved. Gutting speaks of "basic beliefs", Bloor and Barnes of "interests", and I have introduced "utilities", "meta-reasons" and "arational factors". My terms are closer in meaning to Gutting's basic beliefs (if I understand him correctly). The interests of the strong programme operaye outside cases of scientific reasoning and condition that reasoning as "social forces" condition economic developments and social changes. What I have wanted to show is that meta-reasons or utilities operate within scientific reasoning, and furnish one more mark of the value-ladenness of that reasoning. In the context

in the attempt to understand that in some respects scientific work is surely socially conditioned. But such sociological explanation of some features of the scientific enterprise must, I think, take its place *beside* the explanations of intellectual historians that are based on rationality assumptions or meta-reasons. If science is regarded as a problem-solving activity, then there must be some point at which we can say that problems have, or have not, been solved. I tend to think myself that at that point pragmatic considerations come to play the most important role, and thus my own methodological reconstruction of the progressiveness of science ultimately differs from that of both Laudan and Bloor. I have told some parts of that story elsewhere.[32]

---

of attempting to understand internal scientific rationality I am inclined to treat social forces as Kant treated spiritual forces: they lack local presence; they are not transeunt.

[32] "Scientific Progress: the Laudan Manifesto", *Philosophy of the Social Sciences* 9 (1979); and "Methodology and the Functional Identity of Science and Philosophy", *Probabilistic Thinking, Thermodynamics, and the Interaction of the History and Philosophy of Science; Proceedings of the 1978 Pisa Conference on the History and Philosophy of Science*, Vol. II, ed. J. Hintikka, D. Gruender and E. Agazzi (Dordrecht 1981).

CHAPTER 8

KANT'S THEORY OF MUSICAL SOUND: AN EARLY EXERCISE IN COGNITIVE SCIENCE[1]

1. Kant: Hunter of *A Prioris* or Cognitive Scientist?

Kant is well known as the philosopher who spent his life hunting for *a prioris*, philosophically identifiable characteristics of the make-up of human beings. These characteristics are species-universal, and are necessary presuppositions of the possibility of the success of various kinds of cognitive and cultural strategies. Kant bagged some big game. Space, time and the categories are *a priori* conditions of the possibility of human cognition. God, freedom and immortality are *a priori* conditions of the possibility of morality. The *sensus communis* is the *a priori* condition of the possibility of the universalization of judgments of taste. The hard-won trophies are presuppositions of possibilities. Once they were thought to be properties of a universe well ordered by a substantive god. No longer. Now we must look upon them as entrenched contributions of what it is to be human, as preconditions of human potentialities.

It is right and proper to read Kant as the philosophical big-game hunter struggling through the thick jungles of the *a priori*. It is also difficult. An easier way to understand his critical philosophy is to see it as an attempt to offer explanations of various cognitive tasks, where those tasks cannot be comprehended by consulting direct experience alone. Much of our success in coming to know, understand and appreciate depends as much upon the cognitive strategies we bring to bear on these tasks as it does upon the sensory inputs that trigger the attempts. To perceive objects requires that they be ordered in space and time: successful perception presupposes strategies of spatial and temporal orientation. To know objectively that something is or is not the case requires that objects can be fitted

---

1 [Previously published in *Dialogue* 32 (1993): 3-24.]

A shorter version of this paper was delivered as the Stillman Drake Invitational Lecture of the Canadian Society for History and Philosophy of Science, University of Victoria, British Columbia, May 24, 1990. The final version owes much to the excellent discussion by those who attended the session. I owe a special debt to my colleague, Thomas Lennon, for guidance in matters both philosophical and musical.

into our categorial or conceptual structures: successful knowing presupposes strategies of reducing the subjectively idiosyncratic to the objectively public. To appreciate that something is beautiful requires fulfilment of a particular kind of cognitive experience: successful aesthetic appreciation presupposes strategies that can universalize feelings. And so on. In each of the cases mentioned – sense perception, empirical knowing, aesthetic appreciation – Kant thinks he has located an important set of cognitive strategies, the successful deployment of which attains a particular cognitive aim, an aim unrealizable by consulting sensory experience in the raw.

Kant's theory of musical sound is such an attempt to locate human mental strategies employed in the cognitive task of enjoying the hearing of sounds arrayed in certain ways. The theory reveals Kant the cognitive scientist. The form of the explanation he gives for music is exactly like the form of his explanations of other cognitive tasks, for example, correct sense-perception, objective knowing, aesthetic appreciation of visual objects or of literary works. His explanation of the appreciation of music provides a good example of how his general explanatory program is applied. the theory is also of considerable interest in its own right. for one thing, unlike many earlier attempts to explain consonance, Kant's theory is coherent and testable.

2. Kant's Theory of Musical Sound: Sound and Mathematical Form

Kant's theory is presented in brief sections of *Critique of Judgment* and his *Anthropology*.[2] His thoughts about music take form in questioning whether the experience of music is merely an affective reaction to a sensation (an experienced single impression on the eardrum), or to what he calls a "play of impressions" (KU sect. 51, pp. 324-25),[3] by which he means an ordered and regular set of sensations received either successively or simultaneously (melodically or harmonically). This difference is important for Kant. Affective reaction to a sensation is merely a change in my mental state (KRV A20/B34, A320/B376). When, for example, I feel a pain, my pain state replaces whatever other mental state I was in just prior to feeling the pain. In this respect, felt sensations are modifications of states of consciousness. In contrast to this, to react affectively to a regular pattern or play of sensations involves not directly felt sensation, not just change of mental state, but *reflection*. In this sense – and the point is worth emphasis – ordered tones are not merely sensed, they are *judged*, or estimated, in accordance with an apprehension of their *form*. This fact makes the appreciation of music a *cognitive* experience. As

---

[2] I use James Meredith's translation, *Immanuel Kant: Critique of Judgment* (Oxford University Press 1964). I will refer to this work as KU, and will refer to section and page numbers of the Academy Edition, *Kants gesammelte Schriften*. I will refer to the *Critique of Pure Reason* as KVR, and will follow the custom of giving page numbers of the first and second editions.

[3] "Eindrücke", literally "impressions", but meaning, of course, impressions of *sense*.

usual, Kant has moved us to deeper levels of analysis (KU Part 1, sects. 14, 51).[4]

If the play of sense-impressions is suitably ordered in some way (as are the consonant pitches, which have a mathematically expressible order), then, in Kant's language, a given play of this kind is a "formal determination of the unity of a manifold of sensations" (KU sect. 14, p. 224), and can be taken to be intrinsically beautiful. This is also a point worth emphasis. In the first *Critique* Kant taught that space and time, as forms of sensuous intuition, order or formally determine objects given in sensation (he calls them "appearances"). No appearance of a thing outside us can be given except that it is in three-dimensional space and in consecutive time; no appearance of inner sense can be given except that it is in consecutive time. These forms are species-universal and necessary for there to be any coherent experience of sense-given objects at all. They are formal determinations unifying any given sensory manifold.

Because items of sense come to us after one another in an unending succession, any temporal array or manifold of impressions must be unified in order to be rendered as *one* experience. That the moments thus united can be *counted*, can be given a number, unifies them. All temporal arrays that can appear as single experiences must thus be *formally* united in arithmetically representable ways. Similarly, given spatial arrays (which are also in time) must be unified by having the elements in the array related geometrically. Geometrical form unifies spatial order of sensations. The play of sensations occasioned by hearing the consonant intervals that make up a musical whole,[5] Kant claims (*Anthropology*, sect. 70, p. 244), is something analogous to arithmetical and geometrical form. This is no accident. Just as spatially and temporally ordered sensations receive unity through mathematization, so also plays of consonant intervals possess a mathematical form.

To suggest that ordered musical sounds formally determine a manifold of sensations is not, however, to solve the problem of why we appreciate some sets of sounds and not others. After all, most heard sounds are dissonant or unpleasant. We have to school ourselves to be able to tolerate them, and some we can never tolerate. Furthermore, it had been known for many centuries that the consonant pitches can be expressed mathematically as ratios of parts of plucked strings (the discovery is attributed to Pythagoras). Divide a string into two equal parts, pluck both parts at the same time, and we get the consonant called unison, expressed as

---

[4] See also Immanuel Kant, *Anthropology from a Pragmatic Point of View*, translated by Mary J. Gregor (The Hague: Neihoff 1974).

[5] Kant's word for "play" is "Spiel", a word having as many and varied uses as its English counterpart. The choice of words is not, I think, intended to minimize the importance of ordered impressions of sense. Quite the contrary, since the aesthetic reaction is directed to an object *taken to be* freely created (which is part of what Kant means by his expression "purposiveness without purpose"), we are invited to think of the ordered sensations to which we react aesthetically as playful only in the sense of being taken to be freely created. One is not playing a game if one is compelled to do so. I do not react with pleasure to music in the same way in which I react to felt pains. I will return to Kant's important distinction between single sensations and plays of sensation in the discussion of point 3 of his theory, below.

the ratio 1/1. Divide a string again into two equal parts, pluck first a half part, then the entire string, and we get the consonant called octave, expressed as the ratio 1/2. Divide a string at 2/3 of its total length, pluck first the part, then the whole, and we hear the consonant called fifth, expressed as the ratio 2/3. The consonant called fourth is derived by multiplying ratios as follows: 1/2:2/3 = 1/2x3/2 = 3/4. The result is confirmed by experiment. Not only do these alleged Pythagorean findings define consonant sound in experimentally confirmed ways, they permit formulation of a musical scale.

What we might call the "simple arithmetic of the consonances" helps us to appreciate that the basic "atoms" of music do enter into melodic and harmonic forms that, in Kant's phrase (KU sect. 14, p. 244), "formally determine a manifold of sensations" (tones, or more specifically in this case, pitches). But this much about mathematical form in music had been known and accepted for centuries by theorists about music. The question of the complex relationship between mathematical form and the sentiment of pleasure, however, remains open. I am not suggesting that Kant would dispute this contention. It is important to recognize, however, that in inviting us to think of musical form by analogy with the formation of intuitions under space and time (thus bringing musical form into the explicit provisions of his basic epistemology), he seems to be doing nothing more than accepting the simple arithmetic of the consonances.

That no claim stronger than analogy of forms (musical, spatial, temporal) can be made on behalf of Kant's initial insights into plays of sensations ought not to be too troublesome. Two additional points can be made. First, it is perfectly natural and typical for Kant to want to find something about pleasant sound that can be shared, since the universalizability of claims about taste is a defining feature of the aesthetic experience.[6] That ordered sound can be expressed as ratios of sound produced in spatial contexts provides part of the required publicity. In section 4, we will see how important this point is for Kant. Second, to suppose that for Kant one has given an exhaustive explanation of the fit between mathematical form and felt pleasures of hearing by exclusive appeal to the simple arithmetic of the consonances is to ignore his important brief discussion of formal finality or purposiveness in Part 2, sect. 62 of KU. Such finality is presented as objective and intellectual, rather than as subjective and aesthetic. The concept of a circle is formally purposive in that the concept can be used to actualize many circle-like things in many different contexts. One would have to accept that in a similar way the simple arithmetic of the consonances is formally, not aesthetically, purposive. For many different melodic and harmonic plays of sensations are possible, and

---

[6] Kant argues that subjective taste can be universalized by appeal to the *sensus communis*, a public sense, where reflection takes into account the collective reason of mankind (KU, sect. 40, pp. 293ff). In the *Anthropology* he refers to the appreciation of music as a "social pleasure" (sect. 18, p. 155).

each actualizes the simple arithmetic of the consonances.[7] But if the mathematical regularity of musical sounds is merely formally purposive, then it cannot help to activate aesthetic responses to music.

3. Relevant Considerations from Kant's Theory of the Beautiful

We need to outline some details of Kant's aesthetic theory to move further with this exposition. For Kant an object is judged to be beautiful because it is apprehended as an object taken to be purposive. This does not mean that we know anything about what it is *intended* to be; it means only that its *form*, independently of any specific content or of any specific individual human interest, can be judged only with reference to features of its *design*. this is true of objects of nature as well as of works of art. Both beautiful natural objects and beautiful works of art possess *technical* features. They are taken to be free creations, as opposed to being construed as products of mechanical actions. Space and time are *a priori* constraints on sensing, in the respect that all possible human sensation is formed by them. beauties constrain sensing only because the order or regularity of the beautiful object impresses by means of its form, rather than by means of its content, however interesting that might be in other respects.

A beautiful flower and a beautiful painting are spatially constrained designs, but only to the extent that we take the spaced object to be formally determined by its design: the parts "fit", are "right", and are seen to be so independently of any measurements in actual space. Similarly, a beautiful set of tones (spoken, sung or played on an instrument) is a temporally constrained design, but only to the extent that it affects us as a whole: the intervals of time between sounding notes are not attended to, only the total play of sensations is. When we judge an object given in space to be beautiful, in some way we "despatialize" it. so also, when we judge an object given in time to be beautiful, in some way we "detemporalize" it. This is what Kant means by saying that the apprehension of the beautiful is always an apprehension of form, or design. Forms or designs, literally construed, are atemporal and aspatial.[8]

---

7 I am deliberately ignoring tempered or atonal scales, which violate the Pythagorean scale but are nevertheless heard as beautiful. Kant's theory can probably accommodate tempered scales. I had thought that the fact that we aesthetically appreciate atonic music falsified Kant's theory of musical sound. After conversations with my colleague Thomas Lennon, I am now not so sure. More of this below, section 5.

8 I realize that 'detemporalize' and 'despatialize' are mongrel expressions at best. But Kant is dealing here with the aesthetic object, and that object is not the biological flower, not the painted surface, not the heard sounds. These objects can be given straightforward physical explanations, which depend crucially upon determination of geometrical angles and distances, and often also upon measures of time. It is well known that such physical explanations of the beautiful object are unavailing for the purpose of accounting for aesthetic appreciation. apprehensions of beauty are affective states, feelings. This changes the status of the predicates we employ. it seems uncontroversial to say 'this flower has three petals' or 'this sound is intense enough to break glass of a certain strength'. Such predication

Kant seems to realize that the consonances present a perfect example of what he has in mind in his theory of the beautiful. if judgments of taste (reporting experiences of the beautiful) are to be more than merely idiosyncratic reactions to sensory stimuli, they must, Kant thought, be universally communicable; we must be able, at least as a matter of rational principle, to *share* our aesthetic responses. The manifolds of sensation organized in space and time, as we have seen, are all mathematizable; knowledge about them is objective. My *response* to a beautiful set of tones is not objective, but what causes it is: pitches are sometimes consonant, and consonances are also mathematizable. The art of producing a beautiful play of sensations of sound "can only be concerned with the proportion of the different degrees of tension in the sense to which the sensation belongs, i.e., to its tone" (KU sect. 51).

Has Kant taken us beyond earlier theorists, all of whom admit that the consonances are mathematically regular? What the earlier theorists did not appropriately emphasize is that the mathematical form of consonant pitches provides a unified structure, which, like all mathematical structures, can be appreciated by all normal listeners.[9] We do not as yet have all of Kant's story. It will turn out that although he calls the mathematical form of the consonances a *conditio sine qua non* of our being able to apprehend a sequence of musical intervals as one, this form cannot be the cause of our aesthetic appreciation of music; we must look elsewhere for that cause. We have returned to a recurrent problem for theorists about music: what links the mathematical form to the experience of pleasure?[10]

4. The Theory of Musical Sound: Structures of Language

A brief survey of Part 1, sect. 53 of the third *Critique* will give us Kant's answer. Of those art forms most likely to charm us and to stimulate our minds, music ranks second only to poetry. the experience of music is a form of enjoyment rather than of culture, because music speaks through sensuous impulses without concepts, and

---

seems idle when we are describing aesthetically appreciated objects. A statement like 'I appreciate this beautiful porcelain vase because it is delicate (has a fine form, difficult to produce)' seems to me largely meaningless as an aesthetic judgment unless translated to say 'I appreciate this vase delicately'. Strange as that locution may seem, it catches the affective character of the aesthetic experience in a way that is completely lacking in the other, normal statement of predication. aesthetic objects do not have or lack properties. Some tampering with ordinary ways of speech therefore seem innocuous.

9 It is too simple, but nonetheless true, to say that Kant thought that because we are all in principle capable of doing sums, we are all in principle capable of apprehending beauty in the arithmetical structures of music.

10 For a wonderful survey of the problems of theorizing about the consonances, see H. F. Cohen, *Quantifying Music: The Science of Music at the First Stage of the Scientific Revolution, 1580-1650* (Dordrecht: Kluwer 1984).

agitates the mind whether or not it has been schooled in music. Music is charming. Nevertheless, the response to charm, like the apprehension of beauty, is universally communicable.[11] Strange as it may seem, this human capacity to share our responses to beautiful music provides the clue to Kant's proposed solution to the problem of providing causal connections between physical sound and the experience of pleasure when hearing music.

The charm (or beauty) of music, the "art of tone," is a universally communicable affect of consonant sound. Kant's explanation (KU sect. 53, pp. 328-29) is the following:

> Every language has an associated tone suited to its sense. This tone indicates, more or less, a mode in which the speaker is affected, and in turn evokes it in the hearer also, in whom conversely it then also excites the idea which in language is expressed with such a tone. Further, just as modulation is, as it were, a universal language of sensations intelligible to every man, so the art of tone wields the full force of this language wholly on its own account, namely, as a language of the affections, and in this way, according to the law of association, universally communicates the aesthetic ideas that are naturally combined therewith....[T]he form of the arrangement of these sensations (harmony and melody), taking the place of the form of a language, only serves the purpose of giving an expression to the aesthetic idea of an integral whole of an unutterable wealth of thought that fills the measure of a certain theme forming the dominant *affection* of the piece.[12]

Kant goes on to say that this affection arises from response to a

---

[11] Some tensions in Kant's thought about music must be mentioned here. Normally, Kant distinguishes charm from beauty. The object taken to be beautiful is contemplated, reflected upon; both a sensuous and an intellectual act are involved in the aesthetic response, which is uniquely active, combining as it does both the free play of the imagination and the constraints of understanding. That which is charming, on the other hand, repeatedly arouses our attention, but the mind is a passive recipient of the experience of charm. This would seem to entail that music is not beautiful but only charming. I do not think Kant has a consistent view of the difference between charm and beauty. His discussion of pleasure in music appears to depend only on a distinction between passive and active reception of that which is taken to be beautiful. The seeming inconsistency emerges from the fact that he takes music to arise directly from sensation, and not from that kind of mental culture that expands our cognitive capacities. Because music is nonconceptual, it must rank below those arts that are directly expressive of concepts or aesthetic ideas. However, because he thinks music has form to which we respond, and form or design is the hallmark of beauty, he does seem to hold that music can be beautiful, although passively received, whereas other arts – poetry, for example – also require attention to conceptual content. He claims quite explicitly in KU sect. 53 that consonant music gives expression to an aesthetic idea. Surely he holds that charm is compatible with beauty and writes in ways suggestive of the point that aesthetic experiences of what is charming can be had. I will return to this question below in section 4 in my discussion of the crucial passage at KU sect. 51.

[12] In the *Anthropology*, Kant writes as follows (sect. 18, p. 155): "As for vital sense, music, which is a regular play of aural sensations, not only moves it in a way that is indescribably vivacious and varied, but also strengthens it; so music is, as it were, a language of mere sensations (without concepts). Its sounds are *tones*, which are to hearing what colours are to sight – a way of communicating feelings at a distance to all those present anywhere within a certain space, and a social pleasure that is not diminished by the fact that many people participate in it."

"proportion in the accord of the sensations," which can be expressed mathematically as the numerical relation of vibrations of the air at the same time, composing the tones either successively or simultaneously. This mathematical form cannot be conceptualized in a determinate way, but it brings pleasure through reflection on the number of concomitant or consecutive sensations coupled with their play. This form is a "universally valid condition" of the beauty of music, and "it is with reference to it alone that taste can lay claim to a right to anticipate the judgment of every man" (ibid.).

This extraordinarily complex and deep account of how music pleases is stated by Kant in a little less than two pages of text. Let me try to express its message as it pertains to our problem of relating consonance to pleasure in hearing music. Point by point, Kant's theory seems to go something like this:

(1) Every expression in (spoken) language has a tone that is expressive of its *sense*. It is important to note that language conveys meanings in several different ways. A spoken phrase must have grammatical integrity if it is to communicate; in most cases, it will also denote or have reference to an object or state of affairs.[13] Music, Kant claims (correctly), has no denotation. Even the mathematical form of a musical theme has no denotation; it is, as Kant says, conceptually indeterminate. This means that pitches, even pleasing collections of pitches, have in themselves no reference to anything external to their being heard as what they are.

(2) What tones associated with spoken phrases *mean* has to do with what affects in the speaker are being expressed. What is being expressed in spoken words (and this is quite a different dimension of meaning than syntactical fit or reference) has to do with what the speaker *feels*. The intonation contours of spoken phrases are to be understood as being much like facial expressions or other bodily displays of affection.

(3) Common and hence communicable intonation contours or "modulations" constitute a "universal language of sensations" (ibid., p. 328). 'Sensations' is here to be emphasized. Tones of linguistic expressions communicate *sense* as a form of understanding what is being said, not because it is syntactically correct, or is understandable because of familiarity with its denotation, but because it communicates a sensory response to what is being said. We do not learn the semantics of tones in the same way in which we learn the semantics of non-modulated spoken language. In such spoken language, the order of the words and their referents are important because the word order and the referential semantics mirror the order of written language; in the language of sensations, the intensity of the tones is important. Intensity of tones communicates expressed feeling.

For Kant, normal sensations of touch, taste and smell are presumably sensory impulses unmediated by reflection. The three senses do not react to "plays

---

[13] A phrase can have grammatical integrity without being grammatically *correct*. To have grammatical integrity, a phrase must only be recognizable by speakers of the language as an allowable group of sounds or of inscriptions. Much grammatically incorrect intoned spoken language has integrity. Consider the dialects of North American teenagers, who communicate with one another in grammatically incorrect ways, although in ways which for them have integrity.

of sensation" in the same ways in which sight and hearing do. touch, taste and smell do not have associated "languages."[14] This view appears to motivate Kant at KU sect. 42, p. 302, where he writes as follows:

> The charms in natural beauty, which are to be found blended, as it were, so frequently with beauty of form, belong either to the modifications of light (in colouring) or to sound (in tones). For these are the only sensations which permit not merely of a feeling of the senses, but also of reflection upon the form of these modifications of sense, and so embody as it were a language in which nature speaks to us and which has the semblance of a higher meaning.

He goes on to identify each colour of the spectrum with a particular affective state. In his language, each colour "disposes the mind to [an] idea": white/innocence, red/sublimity, orange/courage, yellow/ candor, green/amiability, blue/modesty, indigo/constancy, violet/tenderness.[15] He does not offer a similar translation manual for the consonances. Nevertheless, his analogical comparison of music and colour as languages resulting from reflection on form serves to illuminate his view that

---

14 A view that might well be disputed by the lover of fabrics, the gourmet and the blender of perfumes. I cannot pause to investigate the matter here, but it may be that one can make out a difference between the "languages" of touch, taste and smell, on the one hand, and sight and sound on the other, by pointing out that the first three languages are intrinsically *metaphorical*, whereas the second two are *literal*. This would seem to be a distinction Kant could accept, given that, in the case of colour, specific affective states are the "language" of particular colours. Those who classify wines on the basis of modes of taste and odour, and those who distinguish between textures by refined forms of touching will dispute my suggested distinction between the kinds of "languages" of sensation. The experienced oenophile or the blender of perfumes will point out that the sense of smell can detect 5,000 aromas and flavours. The University of California-Davis Aroma Wheel presents a staggering array of nasally detectable sensations. See Ann Noble, et al., "The U. C. Davis Aroma Wheel", *American Journal of Enology and Viticulture* 35 (1984), pp. 107-109. Whether these modes of the sense of smell have an associated language of sensations is a moot point.

15 The colours of the spectrum. See Newton, *Opticks*, Book 1, Part 2. Kant seems to presuppose, but does not discuss, Newton's confidence that each of the colour bands (from red to violet) can, by analogy, be linked to similar vibration bands, the breadth of the seven colour bands in the spectrum being viewed as analogous to the seven string lengths needed to produce the scale. Newton's analogy is somewhat arbitrary. The colours of the spectrum actually blend imperceptibly into one another, and there is no reason to choose indigo as a colour between blue and violet. It has been pointed out that he might just as well have chosen a colour between blue and green, or between orange and yellow, or between yellow and green. Reliance upon the seven notes in the musical scale no doubt prejudiced Newton's identification. Kant's insistence that the privileged scale is that used for sight-singing (see note 20) must have made it easy for him to accept the Newtonian analogy. (See Percy A. Scholes, *The Oxford Companion to Music*, 10th ed. Oxford: Oxford University Press, 1970, pp. 205-206.) I have not been able to locate the source of Kant's identification of each spectral colour with a particular affective state. He leaves the impression that these identifications are as familiar as the identifications of various objects with the different wedding anniversaries.

musical form mimics intonations of speech which express affection.[16]

Kant's stress upon the analogy between colours and pitches is related in important ways to the question of the distinction between reactions to single sensations (which are simply pleasant or unpleasant, and hence not aesthetically appreciated) and to plays of sensations (which are reflected upon, and hence can give rise to aesthetic experiences). At *KU* sect. 51, pp. 324-25, he tells us that apparently we cannot determine with confidence whether colours or tones are just agreeable sensations, rather than beautiful plays of sensation. Looked at in one way, it would seem that it is only the effect of the vibrations in air (pitches) or in light (colours) that it is evident to the senses, because the time intervals between the very many vibrations are not noticed or estimated. This would suggest that combinations of colours and tones can be agreeable, but not beautiful. Kant thinks, however, that two considerations weaken this conclusion. First, there are the mathematically regular proportions of musical sound, and of our estimates of it – which, by analogy, can also be applied to colour contrasts.[17] Apparently Kant's point is, once more, that mathematical regularity (what I have called the simple arithmetic of the consonances) provides at least part of what is required in order to have the several received sensations *formed*, rendering them fit for aesthetic reflection and judgment as beautiful. Remember that mathematical forms unify sensuous manifolds.

Kant's second consideration seems initially quite curious. He invites us to consider the cases of persons who cannot distinguish between colours, although they otherwise have perfect sight, and of persons who cannot distinguish between tones (pitches), although they otherwise have perfect hearing. The tone-deaf person cannot distinguish between pitches and atonal sound or just plain noise, and often cannot distinguish between (spoken) poetry and (spoken) prose. Similarly, persons who are unreceptive to colour differences cannot distinguish between light and shade.[18] This appears to mean that some persons cannot recognize *tonal structures*

---

16 In *Anthropology* section 71, p. 247, Kant asks why, among the fine (linguistic) arts, poetry is more prized than rhetoric. His answer: "Because it is also music (it can be sung) and tone: a sound that is pleasant in itself, which mere speech is not. Even rhetoric borrows from poetry a sound that approximates tone: *accent*, without which the oration lacks the alternating moments of rest and animation it needs." I call this passage to mind only to reinforce the centrality of the close connection between sound and language that underwrites much of Kant's theory of music.

17 Kant knew about the ocular harpsichord invented by Louis-Bertrand Castel, employing Descartes's theory of light, as modified by Constantijn Huygens. The harpsichord matched the sounding of pitches with displays of analogous colours. Interest in his instrument and in the theories relating sound and colour continued in Germany for some time after Castel's death in 1757. In a letter to Kant of December 13, 1790, Christoph Friedrich hellwag provides an elaborate description of what Castel hoped to achieve with his instrument. I am grateful to my colleague Patrick Maynard for calling Castel's invention to my attention.

18 In a letter (January 3, 1791) in reply to one from Hellwag referred to in note 16, Kant writes that when he wrote this section of *KU*, he had in mind his best friend, the English merchant Joseph Green, who was tone-deaf. Green had learned to play the clavier by note, but could not tell one tune

either of colour or of pitch. Although Kant does not say so, the required inference seems to be that in order to recognize and to appreciate tonalities of either colour or pitch, something more is required than simple reception of sense-impressions appearing consecutively. Within limits, tone-deaf persons hear the same sounds that I do, but they cannot sing tunes, even clumsily. Just like persons with normal hearing, they receive the sounds one by one, but *no group of sounds is musically significant* for them. They lack the requisite processing structures that would enable them to reflect upon given plays of sensations as distinct from mere sensations.

Kant then contends that music and colour combination can be regarded as beautiful rather than merely agreeable, because those of us who can distinguish between different pitches and colours have a definite limit in our ability to perceive a change in quality in the varying intensities in the scales of colour and of tone; similarly, there is a definite limit on the number of the qualities of the varying intensities that can be distinguished *intelligibly*. Why should these two (apparently empirical) limitations incline us to accept that we can make aesthetic judgments on sets of pitches and combinations of colours?

Kant's implicit story may go something like this. If musical sounds are merely agreeable, then in principle it should be possible for me to isolate just those single sounds that so impress me. But in fact I am severely limited in my ability to separate, *in direct perception*, distinct qualities of sense which serve as distinguishing features of various qualities of intensities of colour reception or sound reception. It follows that at some point my attempt to associate my feeling of agreeableness with just *this* qualitative change in my sense-impression experience would become incoherent. One limitation is purely physical: tones with intensities of pitch and force (amplitude) lying beyond the range of the normal musical scales are simply painful, and thus cannot be isolated as single *agreeable* tones. A second limitation is that tones with intensities of pitch and force occurring beyond the range of human hearing simply cannot be identified at all. A third limitation (the one Kant seems principally to have in mind) is that the number of *qualities* of tonal intensity (tone colours or timbres) is infinite, thus making identification of all single agreeable tones impossible *in principle*.[19]

---

from another when played by someone else. Kant takes this to mean that Green could not distinguish pitches from mere noise. Poor Green also could not tell the difference between poetry and prose, and could not read Pope's *Essay on Man* with pleasure because of its forced and unnatural arrangement of syllables (*gezwungese und geschrobene Sylbenstellung*). Kant also writes that he had read somewhere of a family in England whose members could not distinguish between light and shade, and, although they apparently had healthy eyes, saw all objects as if they were the colour of copperplate engravings (*Kants gesammelte Schriften*, Berlin: Königlich-Preussische Akademie der Wissenschaften, 1902, pp. 11, 245).

[19] Kant does not carefully distinguish the modalities of tone: pitch, force and tone colour (Helmholtz's *Klangfarbe*). When he writes about tone, he seems to have principally in mind pitch, where the pitch of a tone is the number of vibrations completed in a given time. What he means by the "intensity" of a sensation (sound) is ambiguous. He can mean either pitch (frequencies of vibrations) or force or amplitude (intensity of oscillations). Technically, following Helmholtz (*On the Sensations*

This is why Kant includes his final point about the limited number of varying colour and tone intensities that can be distinguished intelligibly, that is, can be rendered recognizable and understandable *conceptually*. Here again, no conceptual apparatus designed to classify intensities of sensation can hope to isolate all sounds that can be taken to be agreeable in principle. In his letter to Hellwag, Kant suggests that the "ladder of vocal utterances" (*Stufenleiter der Vocalen*) seems to be a privileged scale of tones because it is the one we all use for sight-singing.[20] However, Guido's scale, although in some sense closest to ordinary song, is by no means the only or the exclusive way of classifying or conceptualizing tones. For example, it does not countenance tempered scales, which are perfectly adequate conceptualizations of sets of sounds that appear pleasant to the listener. Kant's theory limits conceptual scales of tones to intonation contours of human speech, suggesting that there will be some basically limited way of isolating *structured* tonal intensities. The suggestion is that Guido's scale indeed plays that role. The priority Kant gives here to singing is completely compatible with his theory's hypothesis that good musical sound mimics the intonation contours of human speech. Song, at its best, is an aesthetic elaboration on speech. And the qualities of song (and of instrumental music that is parasitic on song) are limited to the affects resulting from the sounding of the notes on the *Stufenleiter*.[21]

From this analysis in section 51, Kant concludes that it is a matter of definition whether we interpret music to be either a beautiful play of sensations or

---

of Tone, 2nd ed., translated by Alexander Ellis, New York: Dover, 1954, Part 1, chapter 1) the quality of a tone is exemplified by notes of the same force and pitch sounded successively on different instruments. Although Kant does not introduce these technical distinctions, what he means by the quality of a tone, as distinct from its intensity (as either force or pitch) seems to be very much what Helmholtz comes to define more precisely in his concept of *Klangfarbe*. Helmholtz comes to agree with Kant that tonal qualities are most richly represented by the human voice, and writes that "human speech employs these very qualitative varieties of tone, in order to distinguish different letters. The different vowels, namely, belong to the class of sustained tones which can be used in music, while the character of consonants mainly depends upon brief and transient noises" (p. 19).

20 *Kants gesammelte Schriften*, Vol. 11, p. 245. The reference is obviously to Guido d'Arezzo's method of solmization, singing the notes of the scale to the monosyllables *ut*, *re*, *mi*, *fa*, *sol*, *la*, *si*, with *do* eventually replacing *ut*. Thomas Lennon has pointed out to me that Kant's theory is supported by the fact that our musical notation derives from the neumes of medieval music. The neumes evolved from the diacritical marks on Greek characters that directed how plainsong was to be sung. An acute accent marked a rising pitch, grave a falling, circumflex a rising followed by a falling.

21 There are other ways of expressing the notes within the octave, but Kant was probably impressed by the similarity between Guido's scale, and that of Newton, a scale of the pattern: C, D, E flat, F, G, A, B flat. if one assumes, as Kant did (following Euler) that both sound and light appear as the result of vibrations of the air, then the vibration times for the intermediate colours of the spectrum discussed above roughly correspond to the periods of vibration of the notes in Newton's scale. Both Castel, and, I think, Kant, were inclined to take this correspondence literally. For Kant, any method of objectifying (mathematizing) *intensive quantities* was a step forward. For more on Kant and the intensive quantities, see Gordon G. Brittan, "Kant's Two Grand Hypotheses", in Robert E. Butts, ed., *Kant's Philosophy of Physical Science* (Dordrecht: Reidel, 1986), pp. 61-94.

an agreeable (charming) sensation. Only on the first interpretation will music qualify as a fine art, because (by implication) only in this sense is our response a reaction to form, rather than to mere intensity of single sense-impressions. After having adduced such powerful arguments in favour of the first interpretation, it is somewhat surprising that Kant reaches this very tentative conclusion, one that can be construed as conventionalizing the sense in which music provides *aesthetic* experiences. The tensions discussed in note 11 are pervasive in Kant's thinking about music. The problem, at least for Kant, is that music, strictly speaking, has no *intellectual* content, as does poetry. Music is an expression of feeling given in sensation. For Kant, this renders music's aesthetic credentials suspect. I have no doubt, however, that his preference was for the first interpretation.

(4) the "art of tone" (harmonic and melodic arrangement of the consonances) is exactly analogous to speech utterance intonations (*KU* sect. 53, p. 328).[22] It also is a language of affections (is expressive of emotional and other affective responses), and communicates the aesthetic ideas of the tonal structure "according to the law of association" (ibid.). There is no mystery here. Kant means that remembered intonation contours of spoken language are here associated with the patterns of the heard sounds; the structures of tone in language are matched with the structures of heard musical sound. The law of association relates remembered past experiences to present ones because of similarities: the experiences are of the same kind. Similarly, remembered intonation contours are related to present heard tonal contours for just the same reason: they are sufficiently similar in *affect* to be experienced as being of the same kind.

(5) the aesthetic ideas expressed by musical phrases "are not concepts or determinate thoughts" (ibid., p. 329), which means, I think, that once again we are to understand that musical phrases have no denotation, no reference to anything we could seek, for example, as a referential meaning of the octave.[23] Rather, we are to understand that the musical arrangement of the sensations, which *takes the place*

---

22 A contemporary cognitive scientist puts it this way: "It is quite difficult to design a reliable voicing decision procedure, or to formulate a reliable algorithm for computing the larynx frequency, which of course determines the subjectively perceived pitch of the voice; but for the moment we will disregard these difficulties and adopt the convenient fiction that each syllable contains a voiced section (usually a vowel, though not always) of a well-defined if varying pitch. One can in fact produce quite a realistic pitch contour for a natural utterance by assigning musical notes to the syllables" (H. Christopher Longuet-Higgins, *Mental Processes: Studies in Cognitive Science*, Cambridge: MIT Press, 1987, p. 282).

23 This is not the same as saying that we cannot *illustrate* the octave. That can be done by sounding it in any of a variety of ways, by using stretched strings, vibrating forks, or by playing a bit of music on a well-tuned instrument. Nevertheless, the octave is not a concept with a referent, there is nothing in the world that is picked out by the word 'octave'. Being an octave is mathematically describable, and is from time to time a sensation. I suppose Galileo would say that independently of being heard, 'octave' is a *mere* name. It is surprising how much of Kant's account of the sensory reception of tones agrees with Galileo's nominalism of sense qualities. See Robert E. Butts, "Tickles, Titillations, and the Wonderful Accidents of Sounds. Galileo and the Consonances", in Victor Coehlo, ed., *Music and Science in the Age of Galileo* (Dordrecht: Kluwer, 1992).

of the affective form of language, expresses an aesthetic idea "of an integral whole of an unutterable wealth of thought that fills the measure of a certain theme forming the dominant *affection* in the piece" (ibid.). the form can of course be rendered in mathematics. I have suggested above that for Kant this concession provides only a necessary, not a sufficient, condition for taking pleasure in structured tones or pitches.

(6) That the consonances are regular in mathematically expressible ways means for Kant only that we can grasp timed intervals *all at once*. If this were not possible, obviously no pleasure could be taken in what he calls the "beautiful play of sensations" (ibid.). If our appreciation of music were limited to individual responses to *single* pitches *unrelated to any others except that they followed one another in time*, we could not be said to have any aesthetic appreciation of music as the play of sensations. What must happen is that the form of consonance detemporalizes successive sensations, a key part of Kant's theory already discussed above.

5. The Merits of Kant's Theory of Musical Sound: Universalizability and Testability

How are we to understand Kant's theory of the reception of the charm and beauty of music? Does he successfully bridge the gap between heard physical sound and the experience of pleasure? We should note first that Kant's theory avoids the conceptual context of earlier theories which seem to require that we distinguish between minds and bodies. These theories then invite us to seek some intermediate part-mind, part-body thing to link them. Instead, Kant encourages us to consider the possible matching of formal cognitive structures *given empirically*. Here is one way of putting this proposal. I am familiar with modes of linguistic meaning that have to do only with the way in which words are intoned. along with the implicit syntax of the language, and with the semantical rules of reference of that language, I also learn an explicit syntax of the shared intonation contours of that language. These contours are, if you like, *premusical* formal structures. In some ways these structures are simple. I know that my mother is angry, not because of *what* she says, but of *how* she says it; I know that she is happy and at ease, not because of *what* she says, but because of *how* she says it. I store all such responses to meaning (to my felt responses to the affects communicated by spoken words) in my memory. My affects match those communicated by my mother: I feel my mother's anger and her happiness.

Sometimes I hear consonant musical sounds in which I take pleasure. Why do I find them pleasurable? First, because (as all earlier theorists acknowledged) they are mathematically regular; if they were not, I would not be able to respond to them in a *single* experience; I could never say, for example, that some performance of Mozart's *35th Symphony*, *as a whole and as just that play of sensations*, is beautiful.[24] Something else is also involved. The form of the music reminds me of a form or of forms of spoken language. Recognition patterns are available to me that select this musical form as just like, or enough like, remembered intonation contours of spoken language. The law of association to which Kant here refers works in two ways.[25] First, I remember the ways in which I have been affected by what people say. I *feel* the way my mother felt when she said this or that. Second, when I hear consonant sounds, the pattern of these sounds is related by memory to experienced intonations or modulations in speech that sound just right. Musical themes relate to remembered pleasant intonation contours because they share the same *affective* structure. Moreover, the play of musical sensations is expressively much richer than modulation (the "language of sensations"), because it deliberately elaborates upon given tonal structures of spoken language, and because its own structures of pleasant affection are mathematically regular.[26]

Kant has accomplished several things with this theory of the reception of music as beautiful. He has located the respect in which the mathematics of consonance is important. Mathematical regularity is a necessary condition for the remembering of associated beat-like structures. He has also located the point at which experience of beautiful music is universal. All of us are subject to formats of affectivity that are shared in numerous ways. He has provided an empirical theory that will stand or fall by reference to facts of the matter. The "transcendental" justification of mathematical form is surely species-universal. No matter what the language, what the forms of culture, a single experience of a complete musical theme cannot be had unless it is regular in mathematically

---

[24] There is a difficult point here that needs further development in another context. To say that I have enjoyed a performance of Mozart's *35th Symphony* certainly is not equivalent to saying that I enjoyed the single and simultaneous tones which *add up to* that performance, from beginning to end. The aesthetic object in music is not the sum of the inscriptions which present it. That object is a complex affective state in the listener or reader.

[25] In *Anthropology* (sect. 31) Kant discusses three capacities of the faculty of sensibility employed in the construction of ideas: imagination (either involuntary [fantasy], or voluntary [composition or fabrication]), association and affinity. The artist engages in all three cognitive tasks; the beholder in the second two. Kant writes: "The law of *association* is this: empirical ideas that have often followed each other produce in us a mental habit such that, when one is produced, this causes the other to arise as well". Kant thinks that this law is at work in matching various structures: a sound felt now that recalls a previous affective state; hearing consonant sounds and recalling previously experienced patterns of modulation.

[26] As are the steps of a marching cadre of soldiers, the even more expressive steps of a company of dancers, the beats of a drum leading mourners to the side of a grave, and many more such regularities and their associated affective states. When Kant refers to music as giving expression to an aesthetic idea of "an integral whole of an unutterable wealth of thought" (*KU* sect. 53, p. 329), he surely has in mind the fact that musical structures call to mind *many and various* resembling structures. In this respect, music is unequalled as an expression of various matching structures that bear definite and remembered affective relationships to one another. He tells us in the letter to Hellwag referred to above that one cannot think of a bit of music without at the same time singing it, even if only awkwardly.

expressible ways. One can ask, however, whether it is true of all music in all cultures that intonation contours of speech (and associated affective modes of expression) come to match tonal structures in the music of that culture. One can also ask, this far into the twentieth century, whether our increasing tolerance of protracted atonality in new forms of musical composition refutes Kant's theory, which requires matching of intonations of speech and other forms of affective expression with consonant forms of the structures of pitch.

When I first wrote this essay, I thought that negative answers had to be given to questions of this kind, and that therefore Kant's theory of musical sound is refuted by two sorts of empirical data: different languages providing different "languages of affection", and the fact of aesthetic appreciation of music that is not an elaboration of intonation contours of human speech. Discussion of the paper after its presentation, and additional discussion with my colleague Thomas Lennon, have led me to reconsider. I now think that Kant's theory can be salvaged, and that it is perhaps the most viable one we have available. I adduce four considerations. First, what we have come to call atonal music likely still preserves the distinction between consonant and dissonant sound that had previously been discussed only in terms of tonal music, at least as closely as does music played on tempered instruments.[27] Second, atonal music is clearly not formless, and is acoustically distinguishable from mere noise. Third, we do not "break out in song" when hearing or thinking about atonal music. Nevertheless, it can be counted as music that elaborates on intonation contours of human speech. The use of *Sprechstimme* (speech voice) in atonal compositions emphasizes a connection with speech.[28] Fourth, although Kant doubtless thought that the affective structures of Indo-European languages were those that had to be matched by the language of music, there is no compelling reason to give priority to any language or family of languages in this regard. Kant's theory is compatible with a relativization of families of languages of affection, each one leading, as we know, to very different idiomatic forms of musical style.

6. The Merits of Kant's Theory of Musical Sound: Styles of Explanation

However one answers such questions, there is, in addition to its testability, a second feature of Kant's theory that is worth our attention. For Kant's theory is of a kind very different from those advanced by earlier theorists. To italicize this point, I will

---

[27] I have not dealt with the fact that the distinction between consonances and dissonances is itself difficult to apply, even to tonal music, because the discussion would take us into details of musicology that are inappropriate here. I do not think that Kant's theory of musical sound needs to depend on this *technical* distinction.

[28] I do not know if Kant was familiar with the operatic recitatives of the eighteenth century, vocal compositions that disregard melody and rhythm and seek to imitate intonations of speech. In all probability he was familiar with the popular eighteenth century *Singspiel*, the forerunner of operetta, a genre combining spoken dialogue, songs and instrumental accompaniments.

briefly compare the theories of musical sound of Galileo, Leibniz and Kant. A theory of the consonances needs to explain the connection between *physical sound*, which consists of vibrations in the air caused by some mechanical means, and the *experience of beautiful sound*, which cannot be caused by solely mechanical means, because the experience itself is not mechanical. Galileo seeks to bridge this ontological gap by means of what I will call a *direct causal* theory. Leibniz's theory is a version of the classical *third man* theory of relating disparate ontological orders. Kant's theory is a theory of *cognitive structures*, structures *of meaning* related to one another by means of psychological processes of "matching".

Galileo's general theory of the reception of sensation teaches that without appropriate sense organs, exterior material stimuli could not cause perceptions of colour, sound and others. In the absence of sensory receptors, colours and words and pains would be "mere names".[29] It is well known that there are severe ontological problems associated with this theory. How can motions of material bodies, causing impacts on sense organs, also cause non-material *appreciations* of certain sensations, and not others? in applying his general theory to the case of the appreciation of musical sound, Galileo seems blissfully unaware of such problems. Briefly, here is what he has to say about musical sound.[30]

The ratio of a musical interval is determined by the number of percussions of air waves on the eardrum, causing it to vibrate with the same frequency. This being granted:

> Agreeable consonances are pairs of tones which strike the ear with a certain regularity; this regularity consists in the fact that the pulses delivered by the two tones, in the same interval of time, shall be commensurable in number, so as not to keep the eardrum in perpetual torment, bending it in two different directions in order to yield to the ever-discordant impulses.[31]

Galileo, faithful to his preferred form of mechanical explanation, gives us a causal account of the interaction of two *physical* motions: vibrations of the air, vibrations of the tympanum. While his theory recognizes the fact that mathematical regularity plays an important role in the reception of beautiful sound, it is incapable of linking mathematical structures, external physical movements, movements of the eardrum and the *experience* of beautiful sound. As one might expect, this explanatory failure

---

29 Galileo Galilei, "The Assayer", in *Discoveries and Opinions of Galileo*, translated with an introduction and notes by Stillman Drake (New York: Doubleday, 1957). For a thorough discussion of Galileo's general theory of sensation as applied to the case of the consonances, see Robert E. Butts, "Tickles, Titillations, and the Wonderful Accidents of Sounds" (note 23 above).

30 Galileo's discussion of music is limited to roughly a dozen pages of the first day of the *Discoursi*. See Galileo's *Dialogues concerning Two New Sciences*, translated by Henry Crew and Alfonso deSalvio (New York: Dover, 1954), pp. 95-108.

31 *Dialogues concerning Two New Sciences*, translated by Crew and deSalvio (New York: Dover, 1954), p. 104.

leads Galileo to discuss the particular consonances not with a scientific, but rather with a quasi-poetic vocabulary. For example: "Thus the effect of the fifth is to produce a tickling of the eardrum such that its softness is modified with sprightliness, giving at the same moment the impression of a gentle kiss and of a bite" (ibid., p. 107). This kind of talk, I submit, helps not at all in the effort to explain the *aesthetic* features of sound reception. Direct causal theories of the reception of agreeable sound seem to presuppose that physics and physiology can exhaustively account for *qualitative* features of sensory experience. That presupposition can surely be disputed.

Unlike Galileo, Leibniz is at least aware that there is a problem of relating physical motions, mathematical structure and appreciation of consonant sound. For Leibniz, all dimensions of *meaning* are functions of the soul. What is needed, then, is something that is part physical, part spiritual, and hence can link vibrations of the air to reactions of the soul. That something, that third man, is the animal spirits. A succinct statement of the theory occurs in Leibniz's well-known *The Principles of Nature and of Grace, Based on Reason*: "Music charms us, although its beauty consists only in the agreement of numbers and in the counting, which we do not perceive but which the soul nevertheless continues to carry out, of the beats or vibrations of sounding bodies which coincide at certain intervals".[32] In his essay "On Wisdom", Leibniz fills in some details of this theory. He first suggests that we do not always observe what constitutes the "perfection of pleasing things", or "what kind of perfection within ourselves they serve". This appears to mean that although we do not understand these matters, we nevertheless feel about them by means of a "sympathy". Furthermore, an account can be given of the causes of what "really appeals to us".[33] He thinks music is a beautiful example of this; we can give an account of why music pleases us, we can solve the problem of the consonances.

Everything that emits sounds gives off invisible impulses which are caused by vibrations or transverse motions ("as we see in strings"). When the sounds are ordered and come with a certain variation, they please us, as do some changes from long to short syllables, coincidences of rhymes in poetry ("which contain a silent music"), drum beats, and the "beat and cadence of the dance".[34] Such regularity is also to be found in the "artfully created" vibrations and beats of

---

[32] *Gottfried Wilhelm Leibniz: Philosophical Papers and Letters*, 2nd ed. Edited, translated and with an Introduction by Leroy Loemker (Dordrecht: Reidel, 1969), p. 641.

[33] *Leibniz: Philosophical Papers and Letters*, 2nd ed., pp. 425-26.

[34] As we have seen in the discussion of Kant, such regular and ordered variations in contexts *other than essentially musical ones* have an important bearing on how we account for the *causes* of the experience of beauty. What all early attempts to account for the consonances simply miss is the point that the causal nexus linking sounds to experiences of sounds is complex and is supplied by *other kinds of experiences*, such as the ones Leibniz lists. Leibniz was within an ace of anticipating Kant's more mature theory. Unfortunately his thought was still burdened by ghosts of the past.

strings, pipes, bells and the air itself. any of these uniformly moving physical objects, when the sound it produces is heard, "creates a sympathetic echo in us, to which our animal spirits respond. This is why music is not sufficiently noticed or sought after" (Leibniz, "On Wisdom").

As is often the case with Leibniz, we are not given very much to go on. As I understand the theory, it involves the following: (1) Sounds are produced by vibrations in the air and (2) are heard (presumably by the physical operations of percussing the tympanum). (3) These heard sounds are fully describable mathematically if they involve appropriate coincidences at regular intervals. (4) We are unaware of a process that involves both the soul and the animal spirits. This process is initiated when sound-generated mathematical structures are taken up by the animal spirits, transported to the soul, allowing creation of the "sympathetic echo in us" (ibid.). That echo (we are unaware of its production from case to case) is our experience of pleasure in the hearing of the consonances. The 'third man' involved in the theory is of course the animal spirits. As animal, the spirits, like physical objects, move. Like the soul, they are a kind of spirit.

Galileo's theory may be overly parsimonious; that of Leibniz is surely too ontologically generous. Galileo was unable to connect physical happenings with subjective events. Leibniz connects them but only by virtue of the postulation of two specious entities: the animal spirits and the soul. What advantages has Kant's theory over these two? First, whereas Galileo fails to relate two mutually irreducible ontological orders, the physical and the mental, and Leibniz manages the relationship only by the introduction of a kind of spiritual causality resulting from the stimulus of physical vibrations of the air, Kant's explanatory strategy does not seem to run afoul of either of these kinds of difficulties. For Kant, there is no question of seeking a physical *cause* of the *subjective experience* of beautiful sound, either in the movements of the eardrum, or in the transportation of the motions of the eardrum by the animal spirits to the soul. Instead, for him it is a question of certain kinds of relations between essentially linguistic structures. *However the sounds are caused*, we possess what he calls the "universal language of sensations", modulation or intonation the "language of affections", which is a mathematically ordered structure of sounds, *mimics* or *models* this language of intonations (*KU* sect. 53, p. 328). The mathematical contours of music are thus pleasant if they mimic or model the pleasant intonational contours of speech. If there is a *cause* of the consonances, it is nothing more than our realized capacity to be affected from time to time by modulation. In this sense Kant's theory seeks to show that consonant sound is a dimension of meaning. The source of that meaning is another kind of human experience – the experience of affection expressed by means to human speech.[35]

[35] Kant's expressive theory of art requires us to accept that artists create what he calls "aesthetic ideas" (*KU* sect. 49, pp. 313-14). An idea, for Kant, has no empirical exemplification or instantiation. In the case of artfully crafted objects, including musical themes, the idea, if you will, delineates the features of the composed design that renders the object beautiful. In the contemplation of such an idea one finds aesthetic feelings of pleasure. What consonant sound, when artfully composed,

Finally, one can measure the strength of Kant's theory along another path. Consider again poor old Joseph Green's tone-deafness. If we grant Kant that Green had otherwise perfect hearing and that the members of the English family had otherwise perfect sight, how would Galileo and Leibniz account for the anomalies of hearing and seeing? Galileo would have to show that the motions of Green's eardrum are defective, or, in a more sophisticated variation, that some other abnormality exists that interrupts transmission of appropriate signals to the brain. This physiological account would provide mechanisms that explain the physical aberrations, but not the subjective experiences. We could still ask why the physical state of affairs prevents *aesthetic experiences* of the requisite kind. Leibniz would have to add some purely ad hoc considerations: either the animal spirits are faulty or the soul is diseased. How one could come to know either of these things would of necessity take us far beyond the limits required for the validation of his theory of musical sound. Metaphysics would have to replace scientific explanation. As we have seen, Kant's suggestion is that certain cognitive structures required for processing qualitative differences in pitch and in colour must be presupposed; otherwise there can be no intelligible distinction between those who can detect such qualitative differences, and those who cannot. This kind of move is typical of Kant, as it is of many contemporary cognitive scientists.[36] The trick is to find what species-universal or humanly invariant cognitive structures make it possible for us to successfully accomplish certain cognitive tasks, in this case, the task of responding with pleasure to variations in pitches or in colours. Interestingly, that there are exceptions to otherwise empirical invariances proves that there are non-empirical invariance-structures. Just as Kant's formalism of space/time and the categories defines scientific sanity, so his theory of musical beauty seeks to define refined taste in beautiful sound.

---

designedly expresses as idea (or meaning) is affection.

[36] Again, Longuet-Higgins, *Mental Processes: Studies in Cognitive Science*, p. 35, with appropriate variations: "We could never hope to understand [aesthetically appreciate] what is said to us [sung for us] unless we were able to use contextual, intonational, syntactic, and semantic clues for decoding the actual sounds themselves". Joseph Green and the members of that English family presumably lacked the capacity to sort out the clues.

CHAPTER 9

KANT'S DIALECTIC AND THE LOGIC OF ILLUSION[1]

> 'A sense of universal illusion ordinarily follows the reading of metaphysics.' Herbert Spencer, *First Principles*, ii, iii, S46, 158 (1875)

In September 21, 1798, Immanuel Kant wrote the following in a letter to Christian Garve: "It was not the investigation of the existence of God, immortality, and so on, but rather the antinomy of pure reason – 'the world has a beginning; it has no beginning, and so on,' right up to the 4th: 'There is freedom in man, versus there is no freedom, but only the necessity of nature' – that is what first aroused me from my dogmatic slumber and drove me to the critique of reason itself, in order to resolve the scandal of apparent contradiction of reason with itself".[2]

In thus proclaiming the centrality of the antinomies for the development of the critical philosophy, Kant invites us to hazard a fresh look at the transcendental dialectic of the *Critique of Pure Reason*. I propose to accept the invitation, and to offer what I hope is at least a partial new understanding of Kant's extraordinary views on dialectical logic.

To begin, it is to be noted that Kant holds a very negative view of dialectical logic as it had been understood and employed by the ancient Greeks. Clearly he is referring to the Sophists, but, given his generally unfavorable opinion of the philosophy of Plato, I think he would have wanted to include him in the general condemnation of those ancients who relied on dialectical reasoning. Kant conceives what he calls '(pure) general logic' as a 'canon' abstracting from all content, and as a science whose operations are governed only by the laws of non-contradiction and excluded middle (and, of course, identity). The ancients, and, I

---

[1] [Previously unpublished]

[2] *Kant: Philosophical Correspondence 1759-99*, edited and translated by Arnulf Zweig, Chicago: University of Chicago Press, 1967, p. 252.

suppose, many others, were tempted to believe that this formal logic could be used to produce new objective knowledge, contentful assertions generated, not by any experiences, but by dialectical logic alone. This belief illicitly converts dialectic into an 'organon'. The conclusions of the application of the logic produce only *illusions*, mere semblances of the truth [A59-62/B83-86].[3]

For this reason, Kant thinks dialectic must be removed from its lofty perch as the highest form of knowledge-production, and demoted to the status of a '*critique of dialectical illusion*' [A60-64/B84-88]. Perhaps the logical indiscretions of the ancients are obvious. Of greater interest is Kant's claim to have discovered dialectical illusion in the arguments of the antinomies.

Kant understood the *modus operandi* of dialectical logic to be *apagogic*: nothing is directly confirmed as true; the truth of a proposition follows inferentially from the falsity of its logical opposite.[4] On this conception of dialectical logic, one who wants to establish that the world began in time first proves that it cannot be true that the world is eternal because the concept of an eternal world is self-contradictory. The trouble is that the logic works in both directions: one who wants to establish that the world is eternal proves that it is a contradiction to regard the world as having begun in time. What matters is that one reduce one's opponents view, by deductive means, to absurdity.

I will not here rehearse in detail Kant's 'solution' to the first two ('mathematical') antinomies. It is enough for present purposes to remember that Kant thought that his conception of transcendental idealism solves the problem of reason's lamentable clash with itself [A490-497/B518-525]. The World, on his view, is not an object existing in itself, it is only represented to us in ongoing experience anchored in sensation. Hence both the thesis and antithesis of each of the first and second antinomies are false. To put it somewhat more accurately: the predicates 'began in time', 'is spatially limited', and 'is made up of simple parts' refer only to temporally transient representations of objects of outer intuition. They are referentially or semantically empty when applied (through illusion) to Worlds. Reason as canonical, perhaps because of the ease with which we can employ it and the regularity with which our best hopes are logically satisfied, fails as an organon, leaving us with the scandal to which Kant refers in the letter to Garve.

Why the scandal? Transcendental idealism, together with Kant's distinctions between the phenomenal and the noumenal, and between the constitutive and regulative employment of ideas, would appear to be all we need in order to avoid embarrassment in the face of the discovery that metaphysics (as dialectic) can no longer be regarded as cognitively respectable (as a source of knowledge) but continues to be essential in its methodological role as seeker after

---

[3] I follow the convention of referring to the first and second editions of *Critique of Pure Reason* as A and B. *Immanuel Kant's Critique of Pure Reason*, translation by Norman Kemp Smith, London: Macmillan, 1950.

[4] Kant had a low opinion of the use of indirect reasoning in philosophy. He thought that, strictly speaking, apagogic reasoning is only legitimately employable in mathematics.

system. Reason may not produce new knowledge, but it does provide the logical structure that all good science seeks to instantiate.

Perhaps Kant's considered view is this one just outlined. Certainly he struggles mightily to sustain the transcendental solution in the Appendix to the Transcendental Dialectic of the first *Critique*, and in the discussion of teleology in the *Critique of Judgment*. These parts of the critical philosophy have received much scholarly discussion. Often his interpreters largely ignore the fact that Kant did, after all, hold that metaphysics is beset by illusions, that metaphysical thinking is delusive. Like it or not, his views on transcendental illusion are crucially central to his entire theory of the philosophical status of metaphysics.

It is one of Kant's most cherished central doctrines that the faculties of the mind be understood as having entirely separate functions. It is the business of sensibility to provide the sensory content of experience without which nothing can be known. It is the function of the understanding to seek to know *about* what is given in sense-contentful experience. Reason supplies the logical form for our science, without which no assurance can be obtained that anything we know is related to anything else that we know. Such a neat compartmentalization is disrupted, however, by the fact that humans judge, make assertions, have beliefs, inclinations, and, sometimes, fall prey to unrepentant stubbornness. Put differently, the objectivity that the faculties of the mind are designed to preserve is often interfered with by our subjective responses to what we are thinking. It is this idea that seems to me to be uppermost in Kant's thought as he discusses the dialectical logic of illusion.

When, for example, I take it as true that the world has a beginning in time, I am, although I may not be aware of it, confronted with an illusion. The "... unobserved influence of sensibility ..." on the objective grounds of my judgment unites with these grounds to produce an error (of judgment) by causing these grounds "... to deviate from their true function"

> just as a body in motion would always of itself continue in a straight line in the same direction, but if influenced by another force acting in another direction starts off into a curvilinear motion. In order to distinguish the specific action of understanding from the force which is intermixed with it, it is necessary to regard the erroneous judgment as the diagonal between two forces – forces which determine the judgment in different directions that enclose, as it were, an angle – and to resolve this composite action into the simple actions of the understanding and of sensibility. [A294-95/B351]

Kant is fond of this way of thinking, introducing what might be called 'the geometry of illusion'. He also employs this geometry in his lectures on logic. For example, in *The Vienna Logic* he is reported to have said:

> When a body is affected by 2 powers, there arises a third movement, where it does not remain on the track of the one or the other, i.e., the diagonal force. E.g., a ball that would have gone directly into the center of the target goes somewhat toward the edge of the target when it is pushed sideways by the wind, and it was

> pushed there neither by the power of the [gun]powder nor of the wind. [824-25][5]

Similarly, the *Dohna-Wundlacken Logic* reports:

> All our errors are turnings of judgments to the diagonal-crosswise (the [German] word [for this] comes perhaps from the English [word] square). [720] [The German for crosswise is *in die Quere*. Square = quadrat][6]

At first glance it seems that what Kant had in mind is the famous parallelogram of forces. A body is carried by an impressed force from A to B in a given time, and from A to C by a second force. Complete the parallelogram ABCD, and by both forces acting together, it will in the same time be carried in the diagonal A to D. However, in dialectical illusion, no body is in motion, and the two 'motions' of the objective content and the subjective principle do not act together to produce the motion along the diagonal. So the "parallelogram of forces" does not accurately represent Kant's intentions.[7]

The question is: what does this quasi-geometry of illusory thought really mean? Just what kind of illusion are we dealing with when we encounter metaphysical notions? Some candidates can be eliminated at once. Metaphysical illusion is not optical illusion, as when I see the straight stick appear to be bent when placed in water. Nor is it moral illusion, which occurs when our best self-interest seems to arise from duty. Nor is it logical illusion (fallacious reasoning arising from inattention to the rules of logic) [*The Vienna Logic* 832; A296/B353]. Kant repeatedly refers to metaphysical illusion as *transcendental illusion* (see A295-97/B352-53), although, as I will try to show shortly, the use of 'transcendental' here is misleading.

Consider three models. Model 1 is the embodied mind as a *tabula rasa.*

---

5 For the lectures on logic I employ the translations of J. Michael Young, *Lectures on Logic: The Cambridge Edition of the Works of Immanuel Kant*, Cambridge: Cambridge University Press, 1992. The references are to the marginal numbers of the Academy Edition of Kant's works, the texts used for translation, and not to page numbers in Young. I think it is highly unlikely that Kant said exactly what is reported in the quoted passage. Consistently with what he says in the first *Critique* and elsewhere, he more probably held that the ball moves crosswise because of the *conjoining* of the two forces, gunpowder and wind. I am grateful to Rolf George for suggesting this line of interpretation, although I do not hold him responsible for my way of developing it. Certainly Kant did not believe that dialectical illusion introduces the action of occult causes.

6 In *The Blomberg Logic* [87], we have: "All errors are, so to speak, crooked lines, which we determine while being driven from the one side by the understanding, from the other side by sensibility".

7 In discussion of the short form of this paper Phillip Cummins suggested that the parallelogram of forces seems better to fit Kant's idea of *normal* cognition, with sensibility and understanding as the two 'forces' which, when they act conjointly, produce knowable objects of possible experience.

But there are no mental blank tablets in Kant's view of knowledge. The chalk of sensation writes upon a tablet possessing the intricate patterns of space and time. Model 2 is normal cognition, where sensibility (pictured as operating from point A to B) and the understanding (operating from A to C) are construed as combining to carry the mental activity from A to D, thus forming the sense-contentful concept. This model, if you like, is the mentalistic equivalent of the parallelogram of forces. Model 3 is a representation of the production of mentalistic illusion, where the motions of understanding and sensibility combine to generate, not an object of possible experience, but a *crooked* diagonal, one that only seems to form a genuine concept, but instead yields a mere semblance of such a concept.

Unfortunately, the models do not really help us to understand just what Kant has in mind when he writes that it is the unnoticed influence of *sensibility* on our attempts to correctly conceptualize that generates the illusions. He repeatedly reminds us that the senses do not err [A293/B350; *The Vienna Logic*, 825; *Dohna-Wundlacken Logic*, 720; *The Jäsche Logic*, 53]. The point is irrelevant here. for there is no proper manner in which I can say that I see, or touch, or smell, or hear a 'world', or any other presumed metaphysical object; that is, if I am talking about sensing as the ordinary having of sensations of things in spatio-temporal contexts. Nor, in an extended meaning of 'sensing', can I be said to have distorted images (metaphysical worlds) of what I actually (in the ordinary meaning of the term) sense.

It is to be regretted that Kant does not show in any detail just what happens when sensibility deceives us. Instead, he refers us to the fact that we judge, thus suggesting that sometimes our judgments are tainted by our inclinations to believe this or that without really having a warrant to do so. What kind of judgments? Certainly not properly formed categorial judgments. Certainly not judgments of taste (aesthetic judgments). And surely not judgments of purpose, either as claims to probability or as guides to practice. Judgments register beliefs. beliefs can be either true or false, and are often based on confusions or illusions. But Kant does not tell us precisely how contaminated judgments can create metaphysical illusion. In other passages, perhaps wisely, Kant abandons the talk about sensibility and judgment and takes a different line.

Reason, in the form of deductive logic, seeks to systematize knowledge by accommodating the results of conceptual knowledge in ever more general, hence simpler, deductive systems. Reason's quest is parsimonious. The old metaphysical adages of parsimony (nature does nothing in vain, nature always takes the shortest path) are replaced by the rational demand to *seek* simplicity, to engage in an endless quest for empirical predicates that are expressively very rich, that, if you will, say a lot. These transformed principles are, Kant thinks, subjectively necessary as providers of logical connections between our otherwise logically quite lonely conceptual claims. The problem arises when we transform the transformation – when we mistakenly construe the subjective principles as objective, thus permitting our crosswise thought to regard metaphysical sentences as making objective claims about things as they are in themselves, as being *about* a world lying beyond the limits of categorized and schematized human experience of things occurring in space and time. This is the kind of illusion Kant calls 'transcendental' [A297/B353-54]. Its creation depends neither on the disruptive influence of sensibility, nor on idiosyncracies of individual judgments.[8]

It is important to note that it is not that any particular (idiosyncratic) subjective response on my part comes to interfere with logic properly applied. A person's like or dislike of temporal beginnings is not at issue. A sudden muscular spasm near my broken femur is not going to help convince me that the world is indeed made up of simple parts. Kant tells us that metaphysical illusion *cannot be prevented.* Just as the sea always appears higher at the horizon than at the shore (since we see it through higher rays of light) and just as the astronomer necessarily sees the moon as larger at its rising, so also we are ceaselessly and unavoidably entrapped by the deceptions of dialectic.

Although Kant, as I have mentioned, calls such deceptive metaphysical principles (revealed in the dialectic) 'transcendental', an additional distinction he marks between *immanent* and *transcendent* principles must be attended to [A295-96/B352-53]. Immanent principles are those "... whose application is confined solely within the limits of possible experience". The principles of the analogies are, for example, immanent principles. Principles whose application seeks to transgress the limits of possible experience, "... to seize possession of an entirely new domain which recognizes no limits of demarcation", are transcendent principles. Transcendental errors seek to apply the categories beyond the limits of experience. Such errors are corrected by a proper understanding of the distinction between the phenomenal and the noumenal, where the noumenal counts as the limit of demarcation. Dialectical errors, on the other hand, are not in this respect transcendental, but are transcendent. Swedenborg's alleged discourses with the

---

8 In the Transcendental Dialectic of the first *Critique* Kant teaches us that the central task of reason is to seek logically simple concepts (he calls them 'conditions') from which less simple concepts may be deductively derived. The ultimate task of *pure* reason is to find 'the unconditioned', that most simple of all concepts from which the entire aggregate of known laws of nature can be deduced, and hence unified and systematized. It is subjectively necessary for reason to seek "to find for the unconditioned knowledge given through the understanding the unconditioned whereby its unity is brought to completion" [A307/B364]. However, "this logical maxim can only become a principle of *pure reason* through our assuming that if the conditioned is given, the whole series of conditions, subordinated to one another – a series which is therefore itself unconditioned – is likewise given, that is, is contained in the object and its connection" [A307/B364]. This (apparently objective) principle is, however, transcendent (see below) in Kant's sense of this term, and can have no empirical employment. To think that it can have such an employment is to be deluded – once again, by the irresistible energy and charm of deductive logic.

Note that although Kant says that the strong form of the principle must be *assumed*, and although his central doctrine appears to be that the assumption is justified in practice, with the principle employed regulatively as a methodological rule, he also at some places suggests that in this form the principle is objective. See A661/B689, where he says that principles of parsimony are "not merely methodological devices". But compare this strange admission with the entire discussion of maxims (methodological rules) in the Appendix to the *Transcendental Dialectic*, a discussion that surely contradicts the slip at A661/B689. Margaret Morrison, "Methodological Rules in Kant's Philosophy of Science", *Kant-Studien*, Heft 2, 1989, pp. 155-172, contains a useful, well-crafted, discussion of the tensions generated by Kant's perhaps inadvertent slides from the subjective to the objective.

angels are in this respect transcendent, determined to confirm the reality of a realm of beings lying outside the limits of (ordinary) human experience. Strictly speaking, such transcendent principles cannot be eliminated. Dialectic can disclose them, and we can then guard against the inclination to take them too seriously, but we cannot get rid of them once and for all. Enthusiasm, *die Schwärmerei*, is a permanent threat to the furtherance of proper philosophical analysis.[9]

One must ponder Kant's claim that transcendent illusion is inevitable. For *all* persons at *all* times? Was there ever a time when Hume construed the principles of parsimony as truths about an experientially inaccessible nature? Do empiricists all begin as committed metaphysicians and only later awaken from their dogmatic naps? Kant's astronomer has a correct theory of the sizes and motions of bodies in our solar system. The theory gives him reasons to allow for the fact that the moon looks so large as 'it rises' even though he 'knows' that the determination of its actual size has nothing to do with this appearance. He will never again be deceived by the appearance of the rising moon – the theory dispels the epistemic influence of the optical illusion. The point is that Kant's critical analysis of metaphysical illusion should give us theory sufficient to dispel the epistemic effects of such illusion, even if the dreams are recurrent and inevitable.

A short form of Kant's story might go like this. We begin our metaphysical speculations with a certain deserved confidence in general logic. Indeed, the *logic* never lets us down. It is our disposition to believe that there are certain objects with knowable characteristics whose ontological status is both atemporal and aspatial that leads us to the conviction that logic, unaided by sense experience, can produce justifiable knowledge. Since there are no such objects, so Kant's story goes, dialectic deceives us. The emphasis here, I firmly believe, should be put upon the fact that those who limit metaphysical results to consequences of an exercise in deductive logic allow themselves to be deceived by that very logic. It is all so easy, and so convincing. Logic is the ground and savior of the a priori. Its results are secure. The highest form of knowledge is *metaphysica ordine geometrico demonstrata*. What happens, however, if the logic, so to speak, turns upon itself, if it allows the deductive proofs of two apparently logically incompatible propositions? At this point we do not dispense with the logic, we insist that dialectic give up its pretensions to being an organon, to possessing a special content.

Very well. Why not, then, simply purge dialectic of the unwanted objects, go on to reformulate metaphysical generalizations as methodological maxims (as Kant insists that we do[10]), and in other respects let the chips fall where they may?

I suggest that Kant gives us at least part of the answer to this question in

---

[9] Those present-day prophets who at the end of the day trumpet the demise of the usefulness of epistemology might do well to ponder this Kantian observation.

[10] See note 8 above, and Robert E. Butts, *Kant and the Double Government Methodology: Supersensibility and Method in Kant's Philosophy of Science*, 2nd edition, Pallas Paperbacks (Dordrecht: Reidel), 1986.

the important Preface to the second edition of *Critique of Pure Reason*. It is not often enough noted that Kant's posture in this Preface is agonistic. The critical philosopher is likened to a policeman [Bxxv]; philosophical works are referred to as "armed" [Bxliv]. What is to be protected, what fought against? Kant tells us: "... speculative reason [must] be deprived of its pretensions to transcendent insight" [Bxxx]. More fully, Kant catalogues the enemies:

> Criticism alone can sever the root of *materialism, fatalism, atheism, free-thinking, fanaticism, and superstition*, which can be injurious universally; as well as of *idealism* and *scepticism*, which are dangerous chiefly to the Schools, and hardly allow of being handed on to the public. [Bxxxiv; Kant's italics]

Of course Kant believes that he can effectuate this botanical murder without loss to the minds of members of the public. They can still believe in God and immortality. The meanings of some of the symbols may have to change, but practical affairs will be unaffected, in fact, greatly improved. The main point is that dialectical logic reveals that not only is metaphysical speculation lacking in cognitive respectability, the belief in the existence of transcendent objects is downright dangerous if allowed to continue unchecked by philosophical criticism. It is for this reason, I believe, that the aged Kant could confidently write (in the letter to Garve) that it was the antinomies that awakened him from (at least one of) his dogmatic naps.

One mild intellectual irritant remains. What are *we* to make of Kant's geometry of dialectical illusion? Is it merely a metaphor, an attempt on Kant's part to show that mathematical pictures, although often not to be taken literally, are nevertheless aids to thought? Or is there something of a more substantial suggestion involved? If so, what does the geometry invite us to conclude, if only hypothetically?

Notice that a subjective principle's interference with an act of objective cognizing (a categorized, schematized conceptualization of something happening in time and perhaps also in space) is itself an *event*. When the lines meet, they point to something happening in me. If this were not so, there would be no point in worrying about the deleterious effects of the free exercise of pure reason. It seems obvious that we must read Kant as having been concerned about the actual cognitive happenings in those 'afflicted' with metaphysics. Metaphysics is epistemologically specious; it is also psycho-physically traumatic.

The diagonal thus represents a datable part of my personal history occurring in me *somewhere*. Where? I think the only place available is my brain. Dialectical illusions, just like optical ones, are neurological events taking place in my brain. In Model 3, the lines moving in different directions and involving incompatible forces, represent paths through nerves, and so does the line expressing the diagonal. If such a diagonalization or crosswise nerve impulse occurred only once or at a few widely separated times, there would, one supposes, be no problem. It is only when the occurrence of that impulse is frequently repeated that trouble arises. If the line representing the diagonal is viewed as frequently reinforced, as making a kind of *groove* in the brain, a person is afflicted with a disease (metaphysics) that, for Kant, needs a cure.

Atheism and fanaticism and the other enemies of criticism involve attitudes and beliefs; they also act as stimuli to behavior. It is not so much that they are false, they are threats to the personal and social moral orders. It is not often enough noted that much of Kant's career was devoted to attempting to prevent the formation of nerve grooves in the brain. He was much concerned with the avoidance of counter-productive reinforcement: his unwillingness to participate in daily religious exercises in the university chapel; his admonition to his students to read less and think more; his (Protestant) objection to lavish ornamentation and monotonous music in churches; his constant warning against memorizing philosophy, because one cannot learn philosophy, but only to philosophize. And of course the touchstone of it all, the message of the essay on enlightenment: to be enlightened is to be free of tutelage, to be the guardian of oneself.

Perhaps in view of these essentially moral considerations we should grant Kant his point about sensibility as what influences us to accept illusion. It could be that he thought not that sensibility *causes* metaphysical illusion, but that the presence of illusion frees sensibility to become *sensuality*. As Foucault[11] notes, Erasmus wrote that madness is that "sweet illusion" that frees the soul from "its painful cares and returns it to the various forms of sensuality". For Kant, such freedom is most definitely not a consummation devoutly to be wished. He was terrified of the potential of the passions of the body to overtake the life of the rational mind.

Still, the groove in the brain corresponds to what Malebranche[12] calls the "sensible emotion of the soul", the passion experienced when animal spirits in the body congregate in one sensory receptor. Foucault correctly notes[13] that there is geometry at work here. The rallying together of the otherwise dispersed animal spirits (or what for Kant is the creation of the groove in the brain) has a form mirrored in the felt emotions of the soul. The motions of the body and the "motions" of the soul are thus reciprocal. Perhaps it is considerations of this kind, involved as they are in the explication of the physiological basis of passion, that we should bring to bear in our attempt to understand Kant's horror of metaphysical illusion. Illusion become passionate – well-grooved – is the basic pathological condition preventing the genuine quest for certifiable knowledge.

Kant may have read François Bossier de Sauvage's *Nosologie méthodique* (Lyon 1772) and would have been much taken with that author's statement that "We call madmen those who are actually deprived of reason or who persist in some notable error; it is this *constant error* of the soul manifest in its imagination, in its

---

11 Michel Foucault, *Madness and Civilization* (New York: Vintage, 1988), pp. 27-28.

12 Nicholas Malebranche, *The search after truth*, translated by Thomas M. Lennon and Paul J. Olscamp (Columbus: Ohio State University Press, 1980), p. 349.

13 *Madness and Civilization*, pp. 134-135.

judgments, and in its desires, which constitutes the characteristic of this category". Kant's groove in the brain is much like the constant error here selected as the deprivation of reason. What makes madness a matter of logic has to do, no doubt, with the fact that a constant error, a well-grooved illusion, can function in the thinking and behaviour of a person within the strictest rules of logical procedure.[14] Although my evidence for this point is circumstantial, it appears at least very plausible that Kant's critique of metaphysical illusion is carried out in the context of well-received contemporary ideas in physiology and in the classification of mental aberrations.[15]

Elsewhere[16] I have tried to demonstrate the great emphasis Kant placed upon physiological and psychological matters of the sort involved here. I regard my reading of the dialectical revelation of transcendent illusion as an additional chapter in what has turned out to be a continuing research project.

---

[14] It is interesting to note that Sauvage contended that diseases of the soul require philosophical persuasion to effect cures. Error must be replaced by truth. Kantss thought holds a mirror to this therapeutic strategy: sound understanding must replace illusion. See Butts, *Kant and the Double Government Methodology*, pp. 298-318, for discussion of Kant's own attempt to develop a psychiatric nosology and his views on philosophy as curative and preventative medicine.

[15] There are many indications that Kant was well-versed in the contemporary literature on psychological and physiological matters. He had read Sauvage, and Cullen, and perhaps Willis. See Butts, *Kant and the Double Government Methodology*, pp. 282-318, for a full discussion of this background to Kant's thought.

[16] *Kant and the Double Government Methodology*.

PART III

TWENTIETH CENTURY PHILOSOPHY OF SCIENCE

CHAPTER 10

THE HYPOTHETICO-DEDUCTIVE MODEL OF SCIENTIFIC THEORIES: A SYMPATHETIC DISCLAIMER[1]

The major philosophical points I wish to make in this essay are not fully original, although I hope that what I have to say makes a coherent and persuasive whole. I shall be discussing a predominating philosophy of science that has held the field against most other systematic philosophies of science for at least five decades. Sometimes this philosophy has been called "Neo-Positivism," (in part to distinguish it from the nineteenth-century positivism of Comte) sometimes, "Logical Empiricism." It shares much with the early "Logical Positivism" of Schlick and other members of the Vienna Circle – a sophisticated philosophical position not to be confused with the positivism of Ayer's shallow *Language, Truth and Logic* – but for many advocates the historical roots lie elsewhere. The labels are not that important, especially since every essential ingredient of the philosophy of science I shall be discussing is to be found in much earlier, diverse philosophies. It is important, however, to pause and consider this historical situation. There are many twentieth-century philosophers who think that philosophy is an autonomous discipline whose history can be ignored; some contemporary philosophers regard philosophy of science as having come into existence only in this century, or maybe in the late nineteenth century. In a typically misleading article on the history of philosophy of science in the prestigious *Encyclopedia of Philosophy*, R. Harré begins philosophy of science in England with the nineteenth-century debate between William Whewell and J.S. Mill on the nature of induction, claiming that it was only then that issues in philosophy of science and general issues in theory of knowledge were separated from one another. This claim totally ignores that for centuries science *was* philosophy, and also that at no time have epistemological issues been independent of those special questions now claimed to lie within the domain of systematic philosophy of science. Aristotle's attempt to account for

[1] [Previously published in *Basic Issues in the Philosophy of Science*, William R. Shea (ed), (New York: Science History Publications, 1976), pp. 36-57.]

causal and demonstrative explanations in physics and biology is surely as relevant to concerns of philosophers of science as is Mill's *Logic*, indeed, probably more so. Perhaps, then, I will be forgiven if I take a somewhat more sweeping and charitable attitude toward both philosophy of science and its genuine historical embodiments.

Developing a philosophical characterization of the nature of science has never been an easy task, partly because of the range and complexity of the problems involved and also partly because the sciences have developed so rapidly that study of relatively stable parts of theories has been impeded. Nevertheless, the history of philosophies of science (at least that section of the history that extends from Galileo to recent logical empiricists) does seem to be represented by a series of views that converge in the twentieth century on a picture of science that might be called the standard hypothetico-deductive model of scientific theory. Any particular philosopher of science selected for discussion would probably differ from some other philosopher on matters of substance both in detail and in programme. If we allow for such programmatic and specialized differences, however, I think that the main features of the hypothetico-deductive model of scientific theories would include at least nine essential ingredients. After developing the model I will attempt to isolate some of its fatal faults.

To begin: On this model a scientific theory is (1) a partially connected set of sentences of different degrees of logical generality. On the basis of this condition, one might refer to the theory as the *sentential* view of scientific theories (which would include, of course, that the *equations* employed in the various sciences be construed as sentences). The view is perfectly straightforward; it is also controversial. As a set of sentences each well-articulated science will contain some generalizations, some empirical claims asserted to hold for all individuals, situations, or events of a certain kind. Some of these universal generalizations will be regarded as *laws*, logically general assertions with counterfactual import, meaning simply that these generalizations will hold (be true) whether or not the individuals or circumstances they talk about continue to be observed. Put differently, science is here regarded as telling us what would happen *if*, and we want to regard its laws as true even if the *if* is seldom actually realized.

Not all universal generalizations will qualify as law-candidates. Thus, "all automobiles have four wheels" needs to be distinguished from "all mammals are warm-blooded." The statement about automobiles is referred to as an *accidental* generalization; if the sentence is true, it is true only because of an accident of history. Most of us would be quite satisfied to call something an automobile if it had fewer or more than four wheels. On the other hand, the statement about mammals makes a stronger assertion. The assertion is not only true (we do observe warm blood in all mammals), but we would want to continue to think that it is true if the species mammalia were selected out of existence. And it would still be true that *if* something came along that was a mammal it would be warm-blooded. It should be noted, however, that although the distinction between accidental and lawlike generalizations is intuitively clear and surely desirable, philosophers have had a notoriously hard time developing the logic of the distinction. (Involved in the question of the counterfactual character of scientific laws is the question of *scientific realism*, a matter of some importance to which I will return later.)

In addition to laws (or lawlike contentions), each such sentential science will contain generalizations of a more impoverished kind, sentences of the form "Under the following very determinate and normally repeatable circumstances, the following event(s) will occur." Sometimes these sentences will be classed as experimental laws or empirical generalizations to contrast them with fully lawlike sentences. They are totally contingent logically, and hence any one of them might turn out to be false. Nevertheless, such generalizations are normally indispensable in specifying the initial conditions under which a fully lawlike generalization can be instantiated and hence subjected to test. At the bottom rung of the ladder each H-D (adopting this rubric as shorthand for hypothetico-deductive) science will contain existential generalizations and statements about actual spatio-temporally fixed events, for example, sentences of the form "There are objects with such-and-such properties," and "At time *t* event *e* took place." These sentences, suitably interpreted for each science, will furnish the principal means for connecting the theory to the "facts" or "data."

(2) The principal machinery employed in stringing together the sentences in the kind of theoretical framework just outlined is the logic of deduction. In the ideal case (rarely, if ever achieved), the theory is fully axiomatized, and each theorem about specific individuals or events will be a logical consequence of the axioms under some suitable empirical interpretation. More usually, the theory will not be fully axiomatized, but an attempt will be made to get as much deductive or calculational strength into it as possible. I said above that an H-D theory is a partially connected set of sentences. I can now be more specific about this caveat: Deductive power is the goal of science; the problem is that a strict deductive connection is not often found to obtain between important sentences in a given science, and if it is found to obtain it is also found to be circumscribed by a set of background sentences (sentences stating initial conditions that instantiate the antecedent of conditional generalizations) that render it somewhat weaker than is normally desired in cases of formally logical deduction. This is true even in those cases where the science is represented mathematically. Very few mathematical "proofs" possess the clean lines of pure deduction. One can understand the initial plausibility associated with wanting scientific systems to have the kind of logical integrity that deductive systems possess. The history of science provides plenty of examples of attempts to get a conceptual picture mirroring a "unified" structure of the world: Newton's inverse square law as a unification of laws apparently so unrelated as Galileo's acceleration-rate law and Kepler's laws of planetary motion; Einstein's attempted unification of relativity theory and quantum theory in a more general "field" theory. The lure of deductive structure is intimately connected with this epistemological ideal of systematic unification of theories. It is also associated with what one might call *pragmatic* dimensions of science, most especially with parsimony principles used as extra-evidential measures of the acceptability of theories-in substantive form: "Nature does nothing in vain"; in evaluative form: "Seek methodological simplicity." I will return to certain additional features of the sentential view of scientific theory that require employment of the deducibility

condition.

In suggesting that a sentential theory contain at least some generalizations which might be laws, I was hinting at what is a most important requirement of such theories: (3) In order not to be regarded as merely empty logical formalisms, they must be "about" something, must be true or false of some "reality" which exists independent of the theory. Seeking to avoid what they regarded as the unfortunate consequences of so-called "subjective idealism" many philosophers have tried to introduce a justified extra-mental or extra-theoretical subject matter that would tell us the way the world is, even if that world should cease to contain knowers. The history of philosophy is full of attempts to establish this external reality. Some thought that we could have direct intellectual intuition of such a reality; others limited their reality to what could be achieved in, or reduced to, the actual having of direct sensations. Kant, thinking that he had discovered epistemological excesses in both of these views, articulated the idea that there is no knowledge without both form (theory) *and* content (that which is independent of theory), thus creating a theory that is in some respects a prototype of the sentential theory we are investigating.

Later, I shall discuss some of the notorious and much-argued difficulties surrounding this concept of the "given." At this point, however, I want to suggest some reasons for the initial plausibility of capturing this elusive philosophical creature. Anyone who pretends to scientific *realism*, anyone who holds that his theories must face the tribunal of independent "facts," will need to think hard about his concept of the given. In the way that C.I. Lewis put it, the idea is quite simple by comparison with more grandiose philosophical ideas: The given is that which we cannot, by taking thought, alter. No amount of thinking will alter *that* I have a pain in my right foot; no amount of thinking will alter *that* I crash into my study wall and am repelled by it if I try to walk through it. Now Carnap and other advocates of the sentential theory sought to capture this feature of "hard" experience by working with the rather more sophisticated concept of an *observation language*, that in some way directly reports observationally inspectable features of the world. In the sciences, in physics for example, this observation language will specify the operational or observational meanings of the terms required in order to *apply* the theories, terms like "straight line," "length," and the whole array of physical and geometrical terms introduced by the theories. Of course, not all of the terms must have such an observational specification of meaning. Physics will continue to contain some purely theoretical terms; it is only required that some of the terms in the theory be linked to observational situations by means of the observation language. If enough of the predictions of a theory thus observationally linked with experience turn out to be correct, then the theoretical terms which are not directly mapped on to experience will be epistemologically justified. We are entitled to think that there are electrons, not because we directly encounter them in our experience, but rather because theories which talk about them turn out to have many correct predictions. (Thus the sentential theory of theories contains an implicit ontological criterion: Something exists – even if that something is only hypothetically postulated – if there are theoretical terms describing it in a theory whose linkage with observation results in correct predictions.)

It is clear that if condition (3) is to be met a further condition must also be satisfied. Condition (4) follows from what I have just been saying: If a theory is to be about something other than itself, the theory cannot be the sole arbiter of meanings of terms. The meanings of terms employed in the observation language must be given independently of the meanings of the theoretical terms as specified in the science. If we can somehow locate an independent given then it will so specify these meanings. If this quest fails, then the best we can do is try various observational and experimental specifications of terms, and then apply the theory under such conditions. At this point the apparently clear distinction between a theory and an observation language begins to become quite murky. For if we are to try various experimental languages, then either we shall just *say* what the terms will mean, or we shall attempt to generate a set of *experimental laws* which will be different from, and have different legal credentials than, the theoretical laws. For example, to learn how to manipulate nature so that we can observe how far objects of differing weights fall in a vacuum is to collect a number of "facts" that give us merely empirical generalizations over the experimental contrivances. But what is the relationship between these experimental "laws" and Galileo's Law, a theoretical law that tells us that in a state of free fall an object moving in the direction of the Earth accelerates at the rate of 32 ft. per second^2? Clearly in order to connect the experimental laws (remember that the meaning of their terms is given without recourse to Galileo's theory) to Galileo's Law we will need an additional law or laws telling us how to "translate" the theoretical statement into the experimental one. Some philosophers have introduced what they call "bridge laws," and "semantical rules" in order to solve this problem. But here I want simply to observe that the problem exists and is a consequence of a seemingly harmless demand that theories be about some "world." In addition, it is interesting to remark that philosophical theories about theories (in this case scientific theories) have the remarkable feature that given sufficient articulation, the philosophical theories begin to generate internal problems interesting in themselves, but that these internal problems take us farther and farther away from our initial task of understanding *something else*.

As serious as this problem of bridge laws may be for some H-D theorists, I am not sure that their account requires this extent of complication. The ontological criterion of the H-D theory allows us to circumvent this problem in the following interesting way: If Galileo's acceleration-rate law is properly formulated in the calculus, then the number of feet that a body falls in so much time (in a vacuum) is a strict *deductive* consequence of the mathematically-formulated law. True, we must stipulate what it means observationally to measure distances and times, but such semantical tampering is relatively harmless, provided only that we have some means of experimentally approximating conditions in a vacuum. Given that this can be done, then any correct prediction of the law will tend to strengthen our confidence that it is right, and each such increase in confidence will bestow some ontological assurance that there are such things as acceleration rates – even though *acceleration rate* is a theoretical construct to which we have no direct means of access by empirical means. At variance here is the H-D ontological criterion and the apparently harmless wish to have theories *directly* tested by means of observation-language claims. Saving the ontological criterion involves in some sense making observation depend upon theory; saving direct observationality involves leaving theoretical terms in some epistemological jeopardy. There is a dilemma here that I shall exploit in my critical remarks about the sentential view of theories.

If, as the above account of the H-D theory entails, each science will have to be linked to a relevant observation language, then it follows (5) that any theories that compete against each other must do so in the same domain of possible facts. And it follows from this condition on competition between scientific theories that each competing theory will contain some terms whose meaning is not at variance with terms in the theory against which it competes: In a given science any allowable theory will have to be so constructed that it contains at least some sentences that imply sentences of the standard observation language "governing" the domain of facts already in some way handled by the earlier theory or theories. Competition between theories – classical mechanics and relativistic mechanics, for example – is thus always to be seen as an attempt to gain explanatory mastery over the same, independently interpreted domain of facts. Theories can change; what theories are about remains constant. For theories to be "about the same thing," they must contain terms which imply common observation sentences in the preferred observation language.

There can be no doubt that characteristic (5) is a consistent consequence of (3) and (4) and that any good H-D theorist must accept it. But the price of this acceptance is high. There are now additional burdens on common observationality and the given. Whatever special problems arose from problems of maintaining conditions (3) and (4), (5), though logically implied by (3) and (4), now places on science a methodological condition difficult, if not impossible, to realize. Many cases are easy: We can perhaps reconstruct classical physics so that both the wave theory of light and the particle theory of light are competitors in the required sense. But an interesting clash between global physical theories like those of Newton and Einstein is difficult to reconstruct in the conditions set by (5). To oversimplify: Relativistic mechanics is not just asking us to replace the old and accepted mechanics of the world of moving objects with a new theory of the way those "objects" move, it is in some sense telling us that the world of moving objects is quite different from the way we thought it was. The conceptual change we are invited to make is just too large to be put off as a competition between two theories for the same ground; it is more like being told that acceptance of a new theory cuts the old ground out from under us (and replaces it with a much less substantial stuff!). Again an obvious ingredient in the H-D theory places it in an unenviably weak position – and the props may by now be in short supply.

Characteristic (6) of the sentential theory of theories has to do with its commitment to a certain structure of explanation, one that is ancient, honoured, plausible, and difficult to expel. Not all positivistic philosophers of science believe that it is the business of science to give explanations. I am here, please bear in mind, giving the most "liberal" reading of the H-D model that I can imagine. And on this model there is agreement with non-H-D theories that perhaps the most central feature of science is that it tries to be explanatory, it tries to tell us what will happen when and if, to tell us what causes what. The most rigorous attempt to expound the conditions of explanation in an H-D context is found in what is now called the "Hempel-Oppenheim covering law model of explanation."

In their classic paper, Hempel and Oppenheim construe explanations as involving a special relationship between sentences of certain kinds. And this identification of explanation with certain logical connections between sentences is, of course, entirely in keeping with the kind of methodology for science that we are investigating. They consider an explanation scheme to involve two parts, an explanandum and an explanans. An explanandum is an empirical sentence describing what it is that requires explanation. To refer to this sentence as "empirical" already prejudices the model, but we will see shortly that the prejudice is modest and easily agreed upon. The explanans is a set of sentences pretending to account for that which the explanandum describes. The question is: Under which conditions can the explanans be said to give such an account? What are the conditions for an adequate explanation? According to advocates of the covering law model of explanation, a proper explanation must satisfy three logical conditions and one empirical condition. The logical conditions are straightforward: First, the explanandum must be a logical consequence of the explanans; the requirement is strict in that one must actually be able to show that the explanandum is deducible from the explanans. (Notice again how easily the idea of deductive connection infiltrates the sentential model of theories.) Second, the explanans must contain at least one general law. Third, at least some sentences in the explanans must have empirical content: In some relevant and specifiable way, the explanans must have something to say about the "world" as we experience it. Finally, a good explanation must meet the empirical condition that the sentences constituting the explanans must be true.

These conditions for an adequate explanation are controversial. Thousands of words have been written in criticism of the model; thousands more in an attempt to vindicate it. Like most philosophers of science working today, I do not find the model totally satisfactory and shall say why below. As with the other features of the H-D theory, however, I think it only fair to point out the strong characteristics of the covering law model, and to stress its initial reasonableness. It has long been thought that the central business of science was to produce laws which explain various natural phenomena. It makes no difference *why* one wants to find these laws. One might be interested in achieving maximal control over nature or other people for reasons good or ill, or one might simply want to know what is lawful about nature. But no such utilities can be attained without knowledge of laws, from which it seems to follow easily that laws do indeed have this centrality in scientific explanation. The Hempel-Oppenheim model tries to capture at least this intuition about the nature of science. So it appears harmless to require that a proper explanation must contain at least one general law, and that this law should be about empirically ascertainable features of the world. Furthermore, it again appears quite harmless to require that the law(s) be true, for it complies with our intuitions about good explanations that what does the explaining (the explanans) not be false or questionable. (There is a problem here: Many are prepared to accept an explanation tentatively if the explanans is only *probable* – quite likely to be true on the basis of *available* evidence. But one must be careful to distinguish between accepting an explanation as true, and accepting an explanation as true until further notice.)

The empirical condition that the sentences constituting the explanans must be true, and the logical condition that the explanans must have empirical content are true for reasons following directly from other features of the H-D model. It is logical that any logically false sentence logically implies every sentence; it is also logical that no sentence lacking empirical content can imply any sentences having empirical content. This is only a problem, of course, if we insist upon the deducibility condition of the model, but then if we were to abandon it, there is no longer any clear sense in which a general law "explains" or subsumes statements about aspects of the behaviour of the world. At least the following benefits accrue from retention of the deducibility condition: Armchair scientists will be prevented from offering as acceptable explanations whose explaining sentences are either false or highly unlikely; and explanations without empirical content will place a heavy burden on their advocates to show us how the empirical world of ordinary experience is related to that which somehow transcends the observable world altogether. It may at once be suggested that the sentential theory of theories thus places "metaphysical" explanations at a disadvantage. Many adherents of this theory would insist that we must find deliberate ways in which this disadvantage be made into an advantage for philosophy of science – by simply abandoning any but scientific explanation. Unfortunately, this seemingly neat move is far from conclusive. One of the major anxieties of H-D theorists in this century has been the problem of banishing metaphysics.

Considerations of this kind lead us naturally to property (7) of the sentential theory of science: Each proper scientific theory will have to be *testable*. This property of scientific theories has been prominent, although not specifically noted, in much of our previous discussion – certainly conditions (3) and (4) have several important points of contact with issues of testability. Here also the theoretical demand appears quite uncontroversial: Everyone knows that it is just this feature of science that sets it apart from other cultural enterprises. The claims of science, so the familiar story goes, can be checked against what we observe either directly or as the result of experimentation. Again, the "facts" determine the truth or falsity of the claims of science. The distressing thing is that scientists and philosophers of science have experienced enormous difficulties in working out a myriad of problems surrounding the concept of a test. It is true that at a certain simple level at which what is at issue is easily agreed upon the results of checking the facts can be relatively decisive. I can fairly easily determine if the Australian rubber tree in my study is sending out a new shoot by entering the study and observing the plant. Of course, we must make modest assumptions about the intelligence of the observer, his adequate eyesight, and other factors. But not much turns on the outcome of this slight challenge, and we are thus prepared to let the "facts" tell us the way it is. But unfortunately the situation in actual science is just not simple in any of these ways.

Certainly testability is not a straightforward matter when it is viewed as a requirement of the sentential theory of theories. As we have seen, a sentential theory is not directly testable by confrontation with experience – many of the terms in such a theory are hypothetical and have no direct empirical counterparts, not even in principle. It is the observation sentences entailed logically by a theory which are subjected to direct test. Positive instances of the observation sentences, when discovered, will give credence to the theory which entails them, but no amount of positive evidence will finally establish a theory as true once and for all. The reason for this is simple and logical: Our scheme is of the form, "If the theory is true, then the following observation sentences will also be true." But we cannot know directly that the theory is true, we can only know that it is true via the truth of its observational consequences. In logic, the inference pattern: If p, then q, q, therefore p, is not universally valid, which means that no amount of confirmation of q will forever establish the truth of p. No matter how strong the evidence is for q, p might still turn out to be false. To use our earlier example: Suppose p to be "All mammals are warm-blooded." q will actually be a set of q's of the form "a is a mammal and is warm-blooded," "b is a mammal and is warm-blooded," etc. Each member of the set of q's will be an instance of p and will confirm p, but no number of q's will establish the truth of p just because it is a universal generalization which might be false even though all known q's are true. Proceeding in the way in which the H-D model must proceed thus generates large problems: H-D theories can be found to be highly confirmed by the evidence, but they can never be established to be true.

Confronted with this unpalatable logico-epistemological result some philosophers simply settled for the weaker situation: We cannot gain scientific truth, but we can turn out attention to the degree in which it is correct to say that one theory is better confirmed than another. (I shall not here enter into the labyrinthine difficulties that have been discovered in the logic of confirmation.) Others – Sir Karl Popper and his followers are the most vigorous defenders of this view – think we should start our reconstruction of the methodology of science from the premise that if we cannot conclusively establish a theory, logic does permit us conclusively to falsify some theories. On this falsificationist way of proceeding, logic is on our side. If the inference scheme: If p, then q, q, therefore p is invalid, the following important deductive inference scheme (called *modus tollens*) is valid: If p, then q, not-q, therefore not-p. A theory T implies its observational consequences E. But if just a single e in the set E turns out not to be the case, then T will be false. It is an important correlate of falsificationist method that one should choose theories that imply a great deal; given large empirical content, they will be easily refuted. Theories with large entailment content which withstand repeated rigorous tests are not shown to be true, they are corroborated in the sense that they have withstood attempts at falsification. There is a certain intuitive merit in falsificationism. The history or science shows fairly decisively that theories once taken to be true are subsequently replaced by other theories. Thus, there is an important historical respect in which science is not getting us closer to the truth by multiplying the number of known-to-be-true theories. On the contrary, history shows inductively that for any given theory at present accepted as true, there are good odds that we will discover later that it is false. Thus the falsificationist view seems closer to the actual movement of the historical epistemology of science by urging us to see the history of science as the history of discarding false theories.

As one might expect, however, things are not all that orderly and joyous in the falsificationist camp. It turns out that whatever problems there are with the logic of confirmation are mirrored in the logic of falsification. Moreover – and this is the only point of logical and epistemological detail that I will introduce – there is a difficulty at the very heart of falsificationism, one pointed out by Pierre Duhem in his influential nineteenth-century book, *The Aim and Structure of Physical Theory*. The problem is that any powerful theory, like the global theories of physics, is not a *single* generalization counting as a theory, it is rather a *set* of interconnected theories plus the background information needed in order to apply such theories. Hence a well-articulated scientific theory faces empirical testing as a body of hypotheses about the world, not as an isolated single hypothesis. The logic of this situation is important enough, I think, to merit a moment's close attention. We have been thinking of a single theory T tested by its set of observational consequences E. Now *modus tollens* tells us that in the simple case where T implies E and E is false, so also is T false. This is a matter of deductive logic: A conditional statement cannot be true if its consequent is false. If T is, however, a set of conjoined Ts, then the logical situation is more complex and needs to be represented as follows:

> (T_1 and ... and T_n) implies E.
> But E is false.
> Therefore (T_1 and ... and T_n) is false.

But if a conjunction of statements is false, then (in the case being considered) either T_1 or ... or T_n is false, but logic alone has no way in which it can tell us how to decide what particular T (or perhaps all of them) is false. It seems literally to be a matter of choice, one that can be made on any grounds whatsoever. From the logic of this situation, Duhem concluded that no single physical hypothesis is testable in isolation. This has come to be known as the "contextualist" thesis, or the "Quine-Duhem" thesis (the American philosopher of science and logician W.V.O. Quine having championed a form of the view with much wider application to all of science). The thesis has obviously powerful and, for some philosophers, unwanted consequences. Certainly, any form of the falsificationist theory of scientific method must face the challenge of contextualism. But here are other difficulties engendered by the thesis; one of them has dominated much discussion of the foundation of physics.

Duhem's contextualism raises the possibility that in the case of important global scientific theories we must acknowledge that the decision between theories can no longer be regarded as a matter decidable simply by putting those theories to empirical test. Something in addition to the "facts" will have to be appealed to in such decisions, a consequence that gives rise to a number of theories of science that argue that, in the absence of hard factual decision procedures, certain extra-evidential measures of the acceptability of theories will have to be appealed to. Kemeny, Goodman, and others suggested that *simplicity* of theories will now come to play its long awaited and justified role in assessing theories. For those scientists and philosophers concerned to maintain favoured theoretical constituents against apparently recalcitrant evidence. Duhem's thesis regally provided a way out which has come to be known as "conventionalism." Poincare and Einstein, both disturbed by the consequences for physics of the new non-Euclidian geometries, appealed to a form of contextualism that puts the challenge of Duhem in sharp relief. Both accepted that no single physical hypothesis is testable in isolation; both drew the conclusion from this that when a physical theory is in empirical trouble, it is the *geometry* that must and can be saved, whatever other parts of the physics must be surrendered in the face of falsifying instances. Both took contextualism as a way of saving Euclidean geometry, but this salvation must be seen as purchased at a very high price. Geometry is critical in the application of physical hypotheses to the world – it provides, when suitably interpreted, a spatial metric by means of which we experimentally measure distances. But if a certain geometry is to have a privileged status in physics, even in the face of falsifying cases, then the observation language of physics is likewise privileged. This will mean, however – recall here the discussion of condition (4) of the sentential model – that the important spatial terms of the observation language are determined by *theory* (in this case, geometry), and hence that the ontology of physics will in large part be fixed, not by successful tests, but by commitment by fiat to a certain theoretical structure. Conventionalism, though forced upon us by Duhem's thesis, leaves us with the stark realization that theory determines observational meaning, that no independent source of the meaning of observation terms is available, and that testing as "consulting the facts" plays a relatively less distinguished role in science than we would have wished.

While much that needs to be said about the H-D model of theories has already been discussed above, two additional features require at least brief mention. The first, condition (8), bespeaks again the extent to which the model appeals to deduction. The H-D theory is committed to *reductionism* in several forms. First, in terms of the model we are required to regard *progression* within a particular science satisfying the model as progression toward greater generality of the laws ingredient in that science. This means that lower-level laws will be "reduced" in the manner of being subsumed under, shown to be implied by, the laws of greater generality. Progression in physics was thought to have been attained when Kepler's laws, Galileo's law, and empirical generalizations describing the behaviour of the tides were all reduced to – thought to be logically derived from – Newton's inverse-square law. William Whewell discussed this phenomenon of reduction in his nineteenth-century *The Philosophy of the Inductive Sciences*, referring to scientific progression as a "consilience of inductions": a convergence of laws derived from different domains by induction upon a single law capable of explaining that such inductive differences can now all be avoided by seeing that the data were after all reports about things of the "same kind."

The second form of reduction is a simple generalization of the first. It may be that some particular science explains, not only the objects and events native to its theoretical involvement, but also objects and events of a different science altogether. What is here displayed in the H-D model is the time-honoured quest for the "unity of the sciences," a unity presumably built upon the foundations of a single foundational science. For some this foundational science is physics, and the many successful attempts to thus "reduce" other sciences to physics has had spectacular results in this century. Both forms of scientific progress through reduction have been much discussed in detail – by Ernest Nagel, for example. But the fundamental point to keep in mind is that the sentential theory of theories continues the age-old "Chinese box" image of scientific progress, where each container box is more general, more "consilient," than the boxes it contains. The "boxes" are, of course, deductively related, and the final greatest achievement of science would be the discovery of the science of all sciences, the generalization that would unify all previous generalizations. Once more it can be said that the writings of Whewell display this image in what is probably its clearest form. One continues to wonder: If deductive structure is weak and wanting, and if direct observationality is plagued by the illnesses I have diagnosed, how, without appeal to convention, can one assess the claims of a pretended reduction, a portended unification? How, that is, within the limits set by the sentential theory of theories?

Lurking in the wings as we have moved along in this discussion of the H-D model – and even muttering an occasional aside – is what I shall arbitrarily assign position (9) in my catalogue of properties of this model. The logically general theories required in the sentential theory are, just because of the requirement of generality, embarrassed by a lack of riches. One must suppose that anyone who asserts a sentence of the form "For all *x* ..." means what he is saying, that is, that his sentence applies to everything in the world, past, present, and future. But direct observation is not this capacious. Among the constraints is the vitally important one that neither I, nor any collection of observers, can penetrate all that is past; neither can we anticipate with completely certifiable foreknowledge what will happen in the future. This is one rather clumsy way of stating what, since the work of Hume, has come to be known as the problem of induction. To raise the discussion one level of technicality we can point out that all logically general statements have reference to all that is, was, and will be; but unfortunately the actually available set of referents of such a universal generalization is observationally finite. Adherents of the sentential theory of theories must, therefore, either offer what they take to be a solution to the problem of induction, or they must presuppose a solution implicit in their philosophy of science. This is not true of falsificationists like Popper, for whom only the dagger of the present case stabbed into the willing – one hopes already dying – body of the theory matters. For any H-D theorist who thinks that the problems of confirmation theory will one day be solved, induction continues its role as the dark, indelible spot, invulnerable to the application of the grues and bleens that would render it biodegradable. (This is a taunting, ingroup, self-indulgent metaphor. Equal time should be given to Nelson Goodman's important and brilliant *Fact, Fiction and Forecast*.)

Throughout my exposition of the nine major characteristics of the sentential theory of theories I have not hesitated to suggest criticisms, nor point up consequences of the theory that may lead to the necessity to abandon it. In the literature in philosophy of science in the twentieth century, the volume of criticism of the model has been enormous, and the types of criticism quite various. No one of these nine points has been neglected. In closing this essay, I want only to introduce two lines of criticism: one as an example of the kind of objection that can be directed at a single vitally important ingredient of the theory; the other not so much a criticism as a sympathetic suggestion of the way in which the theory can be altered to make it acceptable. First, I will turn attention again to the especially important reliance upon deduction characteristic of the model; second, I will try to show that something important and of continuing consequence is involved in the sentential theory's insistence that science must be about a world which it does not create – a world beyond theory.

It must be granted, I think, that in stressing the deductive aspect of science, the sentential theory does indeed correctly mirror a major aspect of the process of scientific inference. As I have mentioned, compliance with the regulative ideal of seeking greater and greater generality appears characteristic of the history of all sciences, and is denied as an ideal only by those for whom – in the most narrow positivistic view – science is a mere description of what has already occurred, a science devoid of genuine explanatory value. Although the sought-after model of scientific inference is deductive, however, the model is seldom historically attained, nor is it free from certain defects.

It was Pierre Duhem who clearly and disturbingly pointed out that the history of science – even the history of physics, where one expects the greatest mathematical precision – yields very few actual cases in which a new general law can be shown logically to imply earlier, less general laws. Indeed, Duhem had the audacity to try to show that not only are some important laws involved in the Newtonian synthesis not deductively implied by Newton's system, but that certain of these laws are logically incompatible with Newton's inverse-square law! (It is not at all easy to show the required contradictions. Leaving aside always troublesome details, the scenario reads something like this: We would need to show that predicted values for various applications of Newton's law differ from those for, say, Galileo's acceleration-rate law to an extent that cannot be accommodated by the calculus of standard observational error and cannot otherwise be standardized in measurable ways by means of non-trivial changes in theory.) Whether or not Duhem is right in this challenge, it is certainly a thesis that cannot be ignored, especially given that historically few, if any, early attempts were made actually to derive Kepler's and Galileo's laws from Newton's law. Only in the subsequent history of physics was the Newtonian synthesis worked out and, in a sense, vindicated. If, in addition to considering laws as mathematical idealizations, we take into account that they are usually interpreted in the context of some general ontology for physical objects, then Duhem's point about logical incompatibility is transformed into one about the difficulty of the *logical application* of idealizations falling short of ontological realization. Thus viewed, Duhem's dagger points at laws like Galileo's and Newton's *within* the bodies of their own systems. Galileo's law requires for its application the existence of vacuua; a firm believer in the view that physical reality is – in its interpretation – particulate, Galileo had to accept a mathematical idealization that contradicted his ontological proclivities. Newton's law is left in something of the same quandary: If the law is true, as everyone believed, then there is action at a distance, a concept clearly at odds with Newton's accepted ontology for mechanics that taught that all physical action is initiated through collision or impact.

One can of course suggest that simple deduction does not depend upon such subtleties of ontological interpretation. And, of course, simply and abstractly it does not. But remember that the sentential theory has locked itself not only into the harness of deductive structure, but also into the required constraints on the *applicability* of a theory, constraints imperatively needed exactly to insure that the deductions – where they obtain – *are not factually èmpty*. Insistence upon the testability of theories and hence upon the requirement that theories have empirical content commits the H-D model to both forms of Duhemian worry about the deducibility requirement.

Some of us might be content to give up the deducibility requirement on sentential (and other) theories on the kind of grounds Duhem supplies. Many philosophers, however, are not readily persuaded by arguments that appeal to historical considerations (just because it is that way doesn't mean that it *should* or *must* be that way), or to historical considerations alone. For the most part, I have been dealing with hypotheses or laws as premisses from which consequences logically follow (or fail to do so). But we also have seen these laws in their role of *consequences* of arguments, as in the cases where *modus tollens* validly applies. In this neighborhood lurks an evil logical genius who cannot, I think, even in the best of seasons be exorcized. That evil genius is the menace of logical deduction running in the wrong direction. We want a body of observational data – expressed sententially in the accepted observation language – that implies a certain hypothesis. But we are clear on the point that such implication cannot be deductive; it can only be confirmatory (we are not here dealing again with Popper's falsificationism, a position taken to have been shown to be inadequate above). But, as we have seen, no body of confirmatory evidence can once and for all establish the truth of its putative law. It follows that any body of observational data can be taken as supporting as many hypotheses as we wish, even in those cases where the hypotheses have different empirical content. Put less esoterically: For any body of observational data O, there is a perhaps infinite set of hypotheses H, any one of which is supported *confirmationally* (notice not *logically*) by O, even if some members of the set H are in the strict sense logically incompatible with one another. This point is not novel; Wesley C. Salmon has investigated this and other shortcomings of the H-D model in his excellent *The Foundations of Scientific*

Inference. For Salmon, the problems are solved by reconstructing the inferential structure of science along Bayesian lines (look this one up for yourself!).

For us, in the present context, the problem is this: If there are residual and damaging problems in Popper's program, and confirmational logical considerations also lead us into trouble, where do we turn? Well, I think the first turn is one that might have suggested itself to us all along: What the sentential theory of theories attempts to do is just not fully possible. On the one hand, this theory strives to characterize science as (ideally – note the normative strain) preserving the central features of deductive systems – science is applied mathematics, at least in the best circumstances. On the other hand, the H-D model insists that the formalism preserving deductive logical connection also must be connected with "reality," where the meaning of this reality is given without the intrusion of theory. I have led you to believe that there are really fundamental problems about the "given," while at the same time trying to argue that some vestige of this idea must be preserved if we are to have sciences whose truth is not simply self-certifying.

The problem here – and I take this to be *the* problem of the sentential theory of theories – is that the sentential theory is the heir of historically determined philosophical *empiricism*. (Of course, this is not going to be the whole story: No contemporary philosophy is so circumscribed that the actual amount it inherits from a certain paternal group can be exactly ascertained.) Something must be said, however, about the links, if not in history, then assuredly in epistemology and ontology. Nevertheless, the damage done by earlier philosophical empiricism to the contemporary sentential theory is not far to seek. The burden of sin of the twentieth-century H-D model of science is that it forces upon us a single ontology that is essentially (in very broad terms) *phenomenalist* in nature. We have already seen part of what this means in our discussion of observation languages. What an observation language gives us is a set of terms whose *meanings* are theory-independent and based on observations. This kind of language, further, is to provide us with *evidence* for or against a certain theory being put to the test. A fascinating and unobtrusive slip occurs: For human beings observation is some kind of seeing (experiencing, in a larger sense). Which means, for openers, that scientific observing must be reducible to seeing, hearing, or otherwise *sensing* – where in the best circumstances, for the sake of the *publicity* of scientific findings, the sensing is done by an *instrument*, a contrivance full of mathematics and packed with assumptions about the nature of human experiences as correlated with certain hypothetical constructs of science. The "blue" that normal human beings see is replaced by correlation with a range of colour bands on the spectrum represented as units of "visible" light, and there are further moves that reduce these intervals on the spectrum to what are called "Ångstrom units," *numbers* of a very special kind. What is going on here is full of significance for the nature of observationality as we must understand it in science.

If the *meaning* of observation terms must finally refer to your sensory experiences and mine, then science is in a shambles of different vivid and dull, sharp and smooth, lively and vapid sense experiences. No generality here, no apparent lawlikeness – most certainly not, if we are constrained to consult the experiences of *everyone*. There is no everyone (note the practice in all sciences of selecting a "sample" from a "population"). The move is obvious: There is no theory-independent observation language, or at least not one whose terms are given a common meaning by the sensory experiences of human subjects. At one extreme (often noted), each experience is given a theoretical colouring by the language by means of which that experience is verbally expressed. At the other extreme – already noted – the experience is correlated to numbers discoverable by all. The point is that in both cases observation is not theory-neutral, but crucially theory-dependent. We arrive at a situation that seems to break the back of the sentential theory. There is no observation language whose terms are given meaning apart from theory; hence there is no "factual" test of a theory; hence two competing theories might be what Feyerabend calls "incommensurable" – sharing no common observation terms, the theories cannot in any logically proper sense be called competitors, even though they pretend to explain the "same" domain of facts. Whewell's dream of consilience is dismissed: The epistemological desire for crucial experiments that will decide once and for all between logically disparate theories remains forever unfulfilled.

Enchanted with the neatness of the logic of crucially deciding experiments, and as yet unaware that Duhem's contextualism was eventually going to come to the forefront of my thought and in some ways dominate it, I reacted against the problem of incommensurability by attacking Feyerabend's early version of this epistemological extreme. I still believe that the motivation behind the criticism was correct – I still want to find satisfactory ways of deciding between competing theories and of accounting for scientific progress. But I am now convinced that part of what Feyerabend attempted in his early work is in principle – if not in detail – correct. I am persuaded that we all operate within the confines of what the late Everett Hall called the "ego-centric predicament": Our givens are all interpreted in some system or other, otherwise we could not understand anything, appreciate anything, love anything, worship anything. But then we have known this, as I suggested earlier in this essay, since Kant; but Kant's arguments are often so obscure that we prefer the simplicity of the sentential theory. In any case, what was right about Feyerabend's early work was his recognition that we need a new concept of evidence that does not rely on empiricist theories of perception if we are to have ways of judging the acceptability of scientific theories.

Feyerabend's general point was that what we need in order to establish the meanings of observational (experiential) terms is not some obscure appeal to common perception, but a pragmatic criterion of the acceptability of certain terms into our language. The insight involved in this "pragmatic theory of observation" is that the kind of evidence we need to test theories (and now we are free to regard theories as being not just sets of sentences, but also policies, ways of acting, etc.) is pragmatic, involving decisions resulting in actions of certain kinds. As a theory of meaning, Feyerabend's account is very defective; but as a suggestive hypothesis about the nature of evidence, it is very rewarding indeed. The sentential theory has done a real service in endeavoring to rescue scientific realism. Feyerabend has shown that what it attempted to rescue was the *wrong kind* of realism. The arguments point conclusively to the conclusion that scientific experience based on a sensory given is mistaken. The "tests" of scientific theories cannot any longer be construed along earlier realist lines. This means, I think, that a radical new thesis must be introduced: What we had earlier regarded as *extra-evidential* measures of the epistemic worth of theories must now come to be recognized as *evidential* measures in the context of a vastly revised notion of what constitutes "evidence."

On this view the simplicity of a theory (not its aesthetic properties, but the number of claims that it can make from few assumptions) will be counted as part of evidence for or against the theory. Likewise, the success of a theory, with regard both to its correct predictions and to how well it *works* when put into practice, will count as evidence. This list of pragmatic-evidential conditions can be extended. The general point is that we come to see that science is a goal-directed activity and, like any such, seeks to achieve the realization of certain utilities. Seen in this light, the question of whether or not our sciences are acceptable, when compared with other forms of activity which seek to realize similar objectives, is answerable by consideration of the results obtained by application of the theory. Another feature of this thesis forces us to view science as a more sophisticated extension of other, more primitive methods of problem-solving. What the earlier American pragmatists would have called the "cash value" of a science is at stake, and, of course, that depends finally upon how much risk-capital one is prepared to venture. The value considerations will, of course, be quite sweeping; no longer will science and value (including ethical value) be seen as distinct – it matters to people whether or not they live in the vicinity of a cyclotron.

What I am sketching as a way out of the difficulties latent in the sentential theory may also give us a pragmatic way of specifying a kind of common context (not language) of observation. A Hopi Indian and I may not speak the same language (and hence may not share the same "theoretical" ontology about the world), but we are, I think, inclined to react in the same way in the presence of an oncoming locomotive, even if the Hopi takes it to be a newly discovered kind of animal and I take it to be a complex piece of machinery. There is this given-in-action that has Lewis's given as its proper ancestor. It may even be that a psycho-physiological account can be given of such reaction patterns. If so, then the "commonness" of experience will again be achieved by means of a correlation between physiological processes and purposive forms of human action. Feyerabend recognized something like this pragmatic dimension of scientific activity.

I have no space left in which to develop my sketch of the new (old) pragmatic realism. A final word must be said, however, on behalf of the sentential theory of theories. It is an important characteristic of philosophical theories of great scope that they are problematic – some might say ambiguous – at just those crucial points where one might hope for the greatest degree of clarity and precision. I take this to be a virtuous shortcoming, or an enlightened vice, of all such theories. The gaps in the arguments of Plato in part made possible the work of Aristotle. The unguarded tenacity of Kepler and Einstein highlighted the problems urgently needing to be solved. It is to the credit of those who worked so hard to preserve various aspects of the sentential theory that they left so much that is unclear and historically inapplicable: They created paths to follow, not sacrosanct dogmas to kneel down to. That is the significance of all philosophy (including philosophy of science): Honestly contemplated, no one knows where a given path will lead. What is important is to choose a path and learn what one can learn by walking it.

BIBLIOGRAPHY

Butts, Robert E., "Feyerabend and the Pragmatic Theory of Observation," *Philosophy of Science*, XXXIII, 4 (1966).

--------, *William Whewell's Theory of Scientific Method*, ed., (Pittsburgh 1968).

Carnap, Rudolf "Testability and Meaning," *Philosophy of Science*, 3 (1936) 4 (1937).

--------, "Foundations of Logic and Mathematics," 1,3, *International Encyclopedia of Unified Science* (Chicago 1939).

Duhem, Pierre, *The Aim and Structure of Physical Theory*, trans. P.P. Wiener (New York 1962).

Feyerabend, Paul, "Problems of Empiricism I," *Beyond the Edge of Certainty*, ed., R. Colodny (Englewood Cliffs, N.J. 1965).

Goodman, Nelson, *Fact, Fiction, and Forecast*, 2nd ed. (Indianapolis 1965).

Grunbaum, Adolf, *Philosophical Problems of Space and Time*, 2nd ed. (Dordrecht, Holland 1973).

Hempel, Carl, *Philosophy of Natural Science*, (Englewood Cliffs, N.J. 1966).

--------, *Aspects of Scientific Explanation and Other Essays in the Philosophy of Science*, (New York & London 1965).

Lewis, C.I., *Mind and the World Order*, (New York 1956).

Nagel, Ernest, *The Structure of Science*, (New York & Burlingame 1961).

Popper, Karl, *The Logic of Scientific Discovery*, (New York 1959).

Quine, W.V., *From a Logical Point of View*, (Cambridge, Mass. 1953).

Salmon, Wesley, "The Foundations of Scientific Inference," *Mind and Cosmos*, ed., R. Colodny (Pittsburgh 1966).

CHAPTER 11

METHODOLOGY AND THE FUNCTIONAL IDENTITY OF SCIENCE AND PHILOSOPHY[1]

I

Scholars put the question expressed by the title of this roundtable discussion [What do history and philosophy of science have to offer one another?] with some urgency nowadays. It is often as if the questioner thinks that the very continuation of the two specializations depended upon an answer to the question *now*. Things do not seem to me to require the sense of urgency, any more than they are made better by one of two relatively – complacent attitudes: "Well, the importance of history to philosophers of science has now been demonstrated, let us turn to questions of what, if anything, philosophy has to offer the historian of science"; or, "Well, we know that all history of science is infested with philosophical considerations, let us turn to the question of whether or not historians have anything relevant to say to philosophers of science." I think it is not often enough appreciated that both history and philosophy of science, as special research and teaching disciplines, are relatively new. After all, it is only in our Century that history of science has become a fully mature and self-conscious technical discipline just as it is only in our Century that philosophy of science, armed with more powerful technology in the form of logic and techniques of analysis, has become, rightly or wrongly, a separable aspect of philosophy.

Earlier in this Century both historians and philosophers of science walked narrow paths. Historical work tended to be locked within the confines of expository

---

1 [Previously published in J. Hintikka, D. Gruender, and E. Agazzi (eds.), *Probabilistic Thinking, Thermodynamics, and the Interaction of the History and Philosophy of Science, Pisa Conference Proceedings*, Vol. II (Dordrecht: Reidel, 1980), pp. 253-270.]

This is a much revised version of the paper communicated to the Pisa conference. I have tried to develop further some of the points discussed at the conference. I also hope that the paper has benefitted from discussions with Jürgen Mittelstrass during a visit to Konstanz following the Pisa meeting. Professor Mittelstrass had just returned from Aarbus, Denmark, where he read a paper entitled "Towards a normative conception of the growth of knowledge" at a meeting sharing common interests with the Pisa conference. It was quite gratifying for me to observe that his study and mine arrive at roughly very similar conclusions, even though our work initially begins from somewhat different orientations.

or narrative accumulation of the details of the past of science, and to building the bibliographical background required to develop this form of history. Philosophers of science, or at least those who turned away from the tradition of combined interest in both history and philosophy of science furthered by the work of scholars like Whewell, Duhem and Peirce, created an almost fraternal model of science as psychology-free, history-free, value-free; free, indeed, of almost everything but the tools of logic and a maturing semantics. How could it have been otherwise?

Faithful to the ordinary demands of customary historiography, the work of early 20th Century historians of science was as it should be: the data needed to be supplied before more enriched speculations about the nature of science could be undertaken. The same kind of professional scholarly constraints (constraints that need not be associated with attendant paranoia) acted upon philosophers of science, and, within the limits of those constraints, the new technical philosophers did what first needed to be done; namely, to attend in earnest to the genuinely philosophical aspects of science: argument and analysis. Historians quite correctly began their inquiry with an attempt to narrate the details of purported contributions to knowledge made by scientists in the past. Philosophers quite rightly sought to understand scientific practice as a complex form of reasoning requiring, before anything else of use could be accomplished, a technical analysis of methodologies and forms of scientific inference. That some of the more breathtaking joys of both history and philosophy were left out of consideration by these earlier researchers was thus necessary for *methodological* reasons. The very nature of their work rendered most historians and philosophers incapable of realizing this cardinal point. Inevitably, the very myopia engendered by the constraints of their own work led each group to begin to mistake the motives of the other.

Thus was created a tension that is with us today, although, hopefully, with diminished effect. It is often the case that different (*different*, not necessarily *competing*) technologies will result in different and, at least, thought-to-be competing ideologies. I suggest that it is the reproduction of this syndrome in the work of early 20th Century historians and philosophers of science that has brought us to the question raised by the title of this session. Historians, in one sense at least rightly confined by their technology and methodology, regarded philosophers as evangelists trumpeting a heretical picture of science as part of a larger world view. Philosophers, in one sense at least rightly questioning the methodological relevance of historical detail to the task of explicating the forms of scientific inference, accused the historians of foisting upon us an ideology of science based upon the commission of the simple logical fallacy of taking the genesis of an idea to be relevant in establishing the validity of that idea. No sensitive scholar in either history or philosophy of science can today be fooled by this syndrome. With hindsight it is just too easy to discern that differences are not necessarily competitions, and that we should learn from this that impaired vision resulting from methodological constraints is, in any and all fields of rational endeavor, subject to the *general* methodological imperative that we should investigate the limits of our own ideology before we accuse others of creating grief through the questionable introduction of their ideologies.

It is precisely in this connection that the work of scholars like Koyré, Burtt, and Crombie is of such extraordinary importance. These thinkers, well versed in history and in philosophy, first pointed up the poverty of applying either the technology of history or the technology of philosophy *exclusively* in an effort to reconstruct a view of the nature of science. Existing, sometimes, quite apart from these methodologies are the richer aspects of the cultural, social, and metaphysical background of science. We need also constantly to remind ourselves of those in both history and philosophy who pursued their work in full awareness of the methodological imperative mentioned above. For example, it was the historian Kuhn who located what might be the limits of history in understanding science. His work on so-called scientific revolutions has repeatedly emphasized the socio-psychological factors that limit science, and in the bargain also limit any narrowly historical or philosophical treatment of the nature of science. On the side of the philosophers, many of those who would be listed as positivists or logical empiricists have, in the investigation of their analytic models, discovered and reported limitations of their models, or have shown willingness to expand their analyses beyond the cramped confines dictated by an earlier application of logic and other analytic tools. There are many examples; for now, think of Goodman and Quine and the enlarged context of pragmatism; or of Hempel and Rudner and the enriched understanding of scientific explanation.

Nevertheless, and this in spite of the happy developments I have mentioned, the parallel progress of history of science and philosophy of science today is not overwhelmingly harmonious. I must here admit to a certain bias, albeit a justifiable one: it is the philosophers and philosophically minded historians who have begun to bridge the gap between the two fields, and it seems to me that this is as it must be. It is up to the philosophers to advance the cause of cooperation. There are, and perhaps I will be excused for pointing this out, some simply *historical* considerations that weigh heavily in favour of this posture. As a professionally organized group, historians of science outrank philosophers of science in the 20th century sense. This has a crucial bearing on the fact that philosophically informed and interested historians of science are really rather hard to find. At least this seems to have been the experience of members of the Joint International Commission on History and Philosophy of Science during the past six or so years. Perhaps this helps to explain why in recent years historical work in logic, methodology and philosophy of science has displayed enormous growth (as evidenced by the number of contributions in these areas made during recent congresses of this Division of the International Union of History and Philosophy of Science); whereas historical accommodations of philosophical interests in science appear to have grown very little (as evidenced by the very skimpy offerings, if any at all, made by philosophers or philosophically oriented historians during recent congresses of the historical Division of the International Union of History and Philosophy of Science). I want now to put forward what may seem a very daring thesis in an effort to provide a framework for understanding this curious situation, and to help in furthering the cause of future cooperation between the history and philosophy of science.

II

The thesis I want to defend, for, indeed, I think it must be defended, would at one time in history have appeared to be a platitude: *there is no interesting and substantial difference between philosophy and science; at least no difference that can mark off the two enterprises as different in kind.* I will call this the thesis of the *functional identity* of science and philosophy. Platitude or not, the message has been forgotten by many recent philosophers, and seems entirely to have escaped the notice of most historians of science. Just consider the following suggestions. What is being done in the intellectual activity of a theoretical physicist is no more, and no less, *speculative* than, say, Wilfrid Sellars's visionary and profound reconstruction of parts of what he calls the "scientific image". Both activities are constrained by a tightly structured rational syntax; both "solve" problems by massive efforts at analysis and systematization; both are controlled by prevailing canons of parsimony; and, *perhaps most importantly*, the *worth* of both activities is determined neither by direct and unambiguous appeal to "facts", nor by considerations of "application" or pragmatic fit, let alone by considerations of direct usefulness. If considerations of this sort were uppermost in the minds of referees of funded research projects, no theoretical work in science, and no philosophical work, would ever be funded – would ever be paid for out of hard coin.

At what is often taken to be another extreme, the importance of the scientific activity is measured by the expectation that results of theory will, at least some of them, be amenable to experiential confrontation. The measure of scientific acceptability is thus in a deep sense epistemological – scientific truths are acceptable just in so far as they reflect accurate experimental or observational results. It is often thought that philosophical speculation or analysis lacks this experiential "hardness". It is argued that what science possesses by way of experiential control is what is precisely lacking in tests of philosophical results. Nothing could be further from the truth. Unless we are to accept the most cynical psychoanalytical view of philosophies as generated *solely* by the unique psychological needs and aspirations of each individual philosopher, I think we must admit that, in some indirect way, philosophies, just like bits of scientific theorizing, must be confronted with some test of experience, where experience, although in some sense theorized, is so minimally theoretical as to constitute something like a publicly available datum.

Of course what we take this or that datum to "mean", what we take to be the "beings" lying behind the data (what are sometimes called our ontological commitments) may vary from individual to individual, culture to culture. All that I am claiming is that the bridging of the cultural or individual gaps does not require any more in the way of theory than is ordinarily needed to "translate" one's experiences into the "language" of others. It must thus be said on Aristotle's behalf that, in the sense of public experience I am suggesting, he, along with any good philosopher, would find a piece of science and a piece of philosophy abhorrent for exactly the same reason if indeed it was in no way amenable to any kind of confrontation with experience.[2] It is important always to guard a fairly hard distinction between ontological (or conceptual) commitments, and truth claims. One is entitled to a private ontology (as in dreams and constructed fantasies); one is never entitled to a private truth. Gnosticism may work as an ontological motivation; it can never work as an epistemologically justified methodology.

Now, if the thesis concerning the historical continuity of speculation, methodology, and the role of experience in science/philosophy and the epistemological barrenness of any attempt to separate the three is correct, as I think it is, then certain interesting consequences flow that touch vitally upon the question embodied in the title of this session. For one thing, if science and philosophy are essentially the same kind of activity, then we can determine at once what it is that historians of science can do for philosophy. In my view, the history of philosophy is simply one part of the history of science (or the history of science is simply one part of the history of philosophy). Thus historians of science have an obligation to work in the history of philosophy; it is part of their field.

Are we onto something important here? There have been very few important philosophers who did not know the history of their discipline. Today, mastery of the history of philosophy by well-trained philosophers is taken for granted. Conceptualization of experience has a history; theoretical explanation has a history. But what about a corresponding cooperative element in the training of historians of science? Has the mastery of the structure of conceptualization, a grasp of the logic of scientific inference, a familiarity with the canons of experiential methodology, been left out – perhaps ignored? Discoveries are not understood if all that we do with them is list them, date them, woodenly correlate them with a man (or woman) and a time. Discoveries are conceptualized expectations; both a logic and a methodology predate each one. Is the historiography of the historian of science too restricted? That, too, is a philosophical/scientific question.

What the historian of science, then, can do for the philosopher of science is exactly the same as what he can do for the scientist – in the best of cases, he can display the rich and complex conceptual, methodological, and experiential components of this or that piece of theory, this or that discovery, this or that failure

---

[2] The reference to Aristotle is a mere historical courtesy. I am aware that the concepts of *experience* I am letting stand without proper explication are full of difficulties that cannot be overcome by simplifications of the sort I introduce here. Concepts of experience were the focal point of discussions in the *first* International Conference on History and Philosophy of Science, with papers by Jürgen Mittelstrass, "Two concepts of experience: Methodological foundations of Aristotelian and Galilean physics", and by me, "Experience and experiment as regulative principles in methodology". Mittelstrass' paper has appeared in German as "Metaphysik der Natur in der Methodologie der Naturwissenschaften", in K. Hubner & A. Menne (eds.), *Natur und Geschichte, X Deutscher Kongress fur Philosophie, Kiel,* 8-12 *Oktober* 1972 (Hamburg 1973). I think we have here a very good example showing why it is better to have proceedings of earlier conferences appear *before* those of later conferences. It helps to keep both the chronology and the scholarship tidy.

in application (or success in application – I am thinking now of the history of technology). One would hope also that such historical work will be deeply evaluative, for one of the things we need to know about science is why, for example, Galilean ways of viewing the world (once these ways are clarified by the application of more refined kinds of mathematics than Galileo himself had to work with) are *better than – are improvements upon* – Aristotelian modes of structuring experience. If we cannot arrive at something like answers to this epistemological/methodological question, there would seem to be little point in regarding the history of science as helpful in understanding science. It is, after all, scientific *success* that we want to emulate, and success is a value (as, of course, is failure). But this gives us an incomplete answer to the question posed by the title of this discussion. We must turn to the question of what it is that the philosopher can do for the historian. This question, as I have already suggested, has a clear logical priority because until it is answered clearly, the historian, I am afraid, can give nothing – nothing, that is, unless one is prepared to settle for a history of science as a history of test tubes and triumphs of discovery, just as some are prepared to accept standard history as a history of beds and battles.

I should say at once that I also think it correct that the philosopher of science can give nothing unless he understands the philosophical/scientific activity as having a history. Indeed, one could argue that the very meaning of his problems is in large part supplied by that history. Can one genuinely be said to understand the quest for charming quarks unless one also understands the kinds of things Thales and Anaximander were seeking to know? Certainly *not* Socrates, who was a social prophet and an unemployed, if very sophisticated, lawyer; it was Plato who made him into a philosopher/scientist. Or, can one genuinely be said to understand the explication of the concept of analyticity without a knowledge of the sources of this concept in mathematics and methodology? Philosophers who diminish history for the sake of incubating logic betray themselves. If what I am saying begins to generate a circle, begins to sound like: one cannot do history of science properly without a knowledge of philosophy/science, and one cannot do philosophy/science properly without a knowledge of history (so that although there is a logical primacy to the question 'what can the philosopher/scientist do for the historian of philosophy/science?' the two questions, in a larger, almost organic context both presuppose answers in the same time; de facto answers, if you like), well then isn't that exactly what we had hoped for – this happy circularity – when we embarked upon the discussion occupying us this past week?

History, just like philosophy in some of its concerns, such as logical analysis and explication of concepts, is without a fixed subject matter, and hence without a standard *general* methodology. Illuminating history, *explanatory* history, therefore presupposes frameworks of interpretation, goal oriented applications, and adaptable methodologies. Thus history, and here it is in positive respects much like the empirically theoretical and methodological aspects of philosophy/science, is not self-critical – its frameworks, applications, and methodologies are imported, extra-historical. There is of course no harm in this, provided that history at some point makes up the balance of payments, and it cannot do this simply by producing more history on the basis of previously imported conceptual and technological considerations. There comes a point at which history ignores its philosophical presuppositions at its peril, for that is tantamount to losing focus, and consequently to losing interest, except for those with the most narrow historical concerns.

No face is lost in this effort at enlightened self-consciousness, for the best strategic models providing frameworks, applications, and methods for history are both philosophically committed and historically biased. I will not here offer argument for this contention, nor will I pretend at this point to be able to offer the historian of philosophy/science a pat presuppositional conceptual background for his work. What I will do is sketch briefly a model for understanding science in a historical context, a model in which various longstanding philosophical problems may someday be seen to be dissipated with increased attention to science as a process and an enterprise that grows and changes over time. I would hope that the details of the model, once worked out, would illuminate the nature of science; for the moment, I offer the model as the basis for conceptual cooperation between historians and philosophers.

III

My model embraces three aspects of the scientific enterprise; for the sake of brevity I will refer to them as the institutional, the methodological, and the epistemological aspects. I do not mean to suggest that science operates in three separate and distinct contexts. Quite to the contrary, the three aspects of science are interconnected, developing, and changing functions of the enterprise; they will be separated here for the sake of the analysis only.[3]

I regard it as a datum that all intellectual quests, including those of the scientist, are undertaken for some purpose. One endeavors to solve an intellectual problem for the satisfaction of some goal that exists independently of the activity of finding the solution. This is true, notice, even if the aim is the satisfaction of curiosity (one may be tempted here to say "mere curiosity", as one often does; but I think the satisfaction of curiosity is a basic human need requiring no external

---

[3] It will be obvious in what follows that I owe many intellectual debts. No one is responsible for my mistakes but me, but I do want to note that my thinking on these topics owes a great deal to what I have learned from Nicholas Rescher, in this connection especially his *Scientific Explanation* (New York & London 1970), and his *Scientific Progress, a Philosophical Essay on the Economics of Research in Natural Science* (Pittsburgh & Blackwell's 1978); from Wesley Salmon, "Bayes's theorem and the history of science", *Historical and Philosophical Perspectives of Science, Minnesota Studies in the Philosophy of Science*, Vol. 5, ed. R.G. Stuewer (Minneapolis 1970): and from Larry Laudan, *Progress and its Problems, Toward a Theory of Scientific Growth* (Berkeley, Los Angeles & London 1977). Salmon offers an important way of conceptualizing scientific success; Rescher notes the absolutely crucial distinction between disciplinary science and knowledge-producing efforts not subject to institutionalization: and Laudan provides the most compelling reading of the *disciplinary success* of science so far available. No one of the three philosophers, however, would necessarily want to put their efforts at the service of my model, and in this respect Laudan will perhaps not be satisfied with my intertwining of various senses of 'success' in what follows.

justification and no defence). The interesting thing to note is that examination of the work of individual scientists does not always, or even often, reveal the aims for the sake of the satisfaction of which the work was undertaken. This is partly because the attempt to satisfy a need often becomes automatic and is, in any case, self-justifying. It is also because the aims of the work do not appear to be essentially involved in the work itself. It is for this reason, I believe, that persons who do highly theoretical and speculative work, work that at least appears to be marvelously abstract and distant from more ordinary human concerns, recoil from the query: "But *why* are you interested in *that* question?", or, "What *good* is a concern for that subject?" The discomfort generated by such questions is at least in part a function of the fact that the questions do not have answers *that are truly required to be given*. After all, the questions are conceptually analogous to: "Why do you eat?", and "Why do you seek to be warm?"; the answers being, obviously: "Because I am hungry", and "Because I am cold". But one is right to be offended by having to give such answers; the offense, after all, is that the question ought not to have been raised in the first place.

As the aims of the work become less prominent – and this seems true no matter what those aims were – the work itself abstracts itself from the continuum of ongoing human concerns, and soon becomes the kind of activity that can be ascertained to be "professional", that is, that can be fraternally protected within the comfortable confines of this or that institution (or discipline, as Rescher would have it). This institutionalization of science results in a network of professionally approved, fraternally accepted ways of engaging in the work of the scientific enterprise. This approval and acceptance, however, is purchased at a high price, for now the institution, operating free of its initial goals, takes on an inevitably political cast: the scientific institution must compete for a place amongst other institutions. Once the preserve of both the wealthy and the ne'er-do-well (the aboriginal amateurs), in its middle period financially advanced by patrons and brilliant dilettantes, the scientific enterprise is now a socio-political competitor for the goods of society, the need for a governmental science policy having become both a moral and an economic necessity. And, given the apparent narrowness of its interests and the esoteric and mysterious nature of its work (remember that the initial aims are now forgotten in the abstraction of the enterprise from those aims), science is a poor institutional competitor. Thus the scientific institution seeks "justification" in engineering applications and in fostering technological advances. In this respect, science betrays its original aims. For politics, as we know, removes the fun from the original activity, and becomes an end in itself. Perhaps this is why politics itself seeks to become a game? After all, one either wins or loses in politics.

It ought not to be thought, however, that the institutionalization of science is entirely disadvantageous. For both the historian and the philosopher of science, the abstraction of science from its initial goals and aspirations has important advantages. For one thing, neither the historian nor the philosopher need seek excuses for the fact that in studying science he deals with an abstraction, for it is not the historian and not the philosopher who creates this abstraction; rather, he simply finds it among the givens of his data. Consequently, the minute investigation of the details of a scientific discovery or the analysis of scientific systems as systems of explanation and confirmation – investigation or analysis uniquely presupposing the availability of detemporalized slices of the scientific enterprise in abstraction – is fully justified by the "facts". This abstract nature of disciplinary science supplies the reason why R.B. Braithwaite could assert in such a blasé and uncommitted manner in his introductory lecture in the undergraduate course in philosophy of science that what he was dealing with is "dead science". It also explains, I think, why the historical curiosity of dating manuscripts (an activity curiously and paradoxically bent upon establishing the *atemporality* of the scientific enterprise), and the absolutely colourless examination of the details of manuscripts (as, for example, in the philosophically vacuous investigations of Stillman Drake), can be so consumingly fascinating as ends in themselves. The point is that all such studies are together justified by the disciplinary or institutional character of research.

These various advantages notwithstanding, the most important benefit of the institutionalization of science comes forward when the detemporalized slices of the activity, historically investigated and philosophically analyzed, are put back together (or, better, when we attempt to put them back together). The reassembly of these dated and conceptualized slices requires a cipher. It is not enough to say – what may be true – that the slices are temporally related, that they follow one another in time. The slices following one another in time are like Hume's sense impressions following one another in time: they are related to one another in no *meaningful* way by being seen to be historically connected. Each detemporalized slice, authenticated and properly dated, comprehended in its logical structures of explanation and confirmation, bears a relationship to other such slices only on the basis of the imposition of the cipher that allows us to put the pieces of the puzzle together. The solving of a puzzle takes place through intervals of time, but the solution is not given by this activity; it is, so to speak, *there all the time*. I would suggest that this cipher is intimately related to the *success* of the science represented in some of the slices or bits of the puzzle, and that the question of success or failure in science drives us immediately beyond questions of the *content* of any or all of the slices into a consideration of the *methods and techniques (technologies)* employed in just those cases that we take to be the best exemplars of scientific work.

Larry Laudan has argued persuasively that it is not the philosophers, but the scientists themselves, who first and foremost articulated the most interesting and successful methodologies.[4] I am not inclined to accept this argument without modification to accommodate my view of the identification of philosophy and science, but in dealing with at least some of the abstractions presented by

---

[4] Larry Laudan, "The sources of modern methodology", R.E. Butts & J. Hintikka (eds.), *Historical and Philosophical Dimensions of Logic, Methodology and Philosophy of Science, Part Four of the Proceedings of the 5th International Congress of Logic, Methodology and Philosophy of Science*, London, Ontario, Canada 1975 (Dordrecht 1977). The University of Western Ontario Series in Philosophy of Science, Vol. 12.

disciplinary science, the point of his argument is entirely correct. In any case, Laudan and I are as one in thinking that the study of methodology is the most promising bridge for relating the efforts of historians and philosophers. To cite a case in which I have myself been at least marginally involved, look at the recent work on Galileo by both historians and philosophers of science. Studies of the detemporalized slices continue: not all of the authentication and dating of manuscripts has been accomplished, the logic has not been fully understood, but enough of the science of Galileo is accessible to provide new insights into how the various Galileo slices fit together. The insights, of course, are insights into Galileo's methods. One would hope that it is not accidental that recent work by both historians and philosophers has converged upon an attempted reconstruction of Galilean methodology. There are many reasons for this convergence to be occurring now, nevertheless, one hopes that all the work being done is based upon accepting that Galileo went about his business in – if not the right way – a way holding out great promise of success in certifying truth claims about the world (methodology is not philosophically neutral; the epistemological presuppositions and consequences are there for all to see). Such acceptance of Galileo's work is by no means gratuitous or easily forthcoming. It must be noted that for all of the fanfare and fuss about Galileo as the father of modern science, his individual contributions to science were not all that many. He did invent a new telescope; he did "discover" the times-squared law. Much of what he did was issue promises for the future. The promise of the unification of physics, a unification he did not himself achieve; the promise of a confirmed theory of the tides, a confirmation he himself failed to give.

The historical and philosophical quest to discover and to understand his methodology must be based, then, not on the overwhelmingly large corpus of new discoveries made by Galileo (these were few), but rather on the confidence we want to bestow on what we take to be his having done science in the "right" way. The elegant and detailed historical investigations of Wisan and Shea and others; the careful and conservative philosophical analyses of McMullin and Wallace and others all seem to point to this way. Even Feyerabend, who thinks that Galileo was wrong on almost every count, nevertheless commits himself to the view that Galileo was one of the greatest propagandists on behalf of science ever to come forward, and he encourages us to imitate his ways of proceeding.[5]

Similar kinds of attention to the revelations of scientific methodology cannot fail to teach us humility. *For even if every putatively positive result of science is in fact negative – false – science is at least successful in that it knows the difference, and can show it to all who will listen attentively, learn actively, and observe with caution and with controlled vision.* From this perspective,

---

[5] For recent work on Galileo, see, for example, the relevant papers in R.E. Butts & J. C. Pitt, (eds.), *New Perspectives on Galileo, Papers Deriving from and Related to a Workshop on Galileo held at Virginia Polytechnic Institute and State University*, 1975 (Dordrecht 1978). The University of Western Ontario Series in Philosophy of Science, Vol. 14; for P.K. Feyerabend see his *Against Method, Outline of an Anarchistic Theory of Knowledge* (London, 1975).

methodology is the sought-after cipher: the pieces of the puzzle do not fit together because of a common shared content, but because of an overlap in techniques, ways of going about doing the business of science. It is not *what* science says that counts; it is rather, that what science says can be known to be either true or false that counts. Thus it is that an enlarged perspective on methodology reveals the often hidden epistemological well-springs of scientific work. Science pretends to tell us the truth about the world, pretends to vindicate Whewell's claim that "Man is the Interpreter of Nature, Science, the right interpretation."[6]

The epistemological questions to which we are led by a study of methodology are essentially questions of *appraisal*, questions having to do with why it is that, for the time being, we will accept the results of a given piece of scientific work. After all, there are plenty of reasons why we might expect a given result to turn out false; indeed, if we believe in induction as applied to the history of scientific discoveries, there is a high probability that it will turn out to be false. I need not in this place rehearse the view put forward by many that it is this very epistemological fragility of scientific results that argues for the health of science as a truth-telling institution. Wesley Salmon has recently argued that the historical success of certain kinds of hypotheses put forward by earlier scientists enters crucially into the probability of the success of a newly introduced hypothesis, by giving us a fix on the prior probability of the new hypothesis based on the past success of other hypotheses of the same kind – a success that yields plausibility to the positive outcome of the new hypothesis.[7] There is an ambiguity in Salmon's claim, a harmless and revealing ambiguity. "Hypotheses of the same kind" can mean many different things. Let me sort out a few of the possibilities. It can mean (1) hypotheses with the same ontological commitments, or, (2) hypotheses with the same *a posteriori* probability or (3) hypotheses introduced in the context of a methodology known to permit truth-value determinations with empirical integrity.

---

6 As I shall be trying to argue in what follows, the standard epistemic predicates 'true' and 'false' lose their integrity when epistemology is thought of as intimately related to methodology. If, with Laudan and others, we take the decisive unit of scientific success to be the problem, then perhaps 'solved' and 'not-solved' will have to replace the standard epistemic predicates. It is curious to note that the epistemic predicates appear to take precedence over other utilities in relatively methodology-free philosophies of science. This seems especially true of the philosophy of science traditionally associated with Hume, and of recent positivism. I have argued that it is appropriate thus to disassociate science from its methodologies for purposes of studying the logical features of scientific systems. But this disciplinary abstraction *is not science*, if what we mean by 'science' has reference to an ongoing historical enterprise. I would suggest that the larger the methodological perspective becomes, the smaller is the true/false content of science seen to be. In any case, it seems to me that the cardinal emphasis in science is not upon true/false decisions *per se*, but simply upon the fact that scientific outcomes provide decisions of some sort. And it is the methodologies, not the abstract (reconstructed) epistemologies, that tell us what will count as a decision.

7 See the essay by Salmon in note 3 above. I have employed Salmon's thesis in another context in my "Consilience of induction and the problem of conceptual change in science", in R.G. Colodny (ed.), *Logic, Laws & Life, Some Philosophical Complications* (Pittsburgh 1977); University of Pittsburgh Series in the Philosophy of Science, Vol. 6.

Any of these hypotheses would satisfy Salmon's condition that the prior probability of a hypothesis is a relative frequency, the relative frequency of the success of hypotheses of the same kind. I will not detail the claim here, but it does seem to me that Salmon himself writes of hypotheses in all three of the above-mentioned senses.

What matters most is how we construe "success" in past hypothesizing and theorizing. Thus, although it is useful and perhaps illuminating to point out the three senses of "sameness", it is more to the point to see that the three senses of sameness are all of them required in order to begin to explicate what we mean by the success of a scientific hypothesis. Ontological commitments make a difference. It is important to note that we have more confidence in hypotheses that explain the motion of objects in this room in terms of commitment to physical objects, measurable distances and times, and calculable velocities and rates of acceleration, than we do in explanations committed to (as in the case of purported poltergeist activity) spiritual beings, nonmeasurable astral distances and times, noncalculable aspatial influences, and the like. Similarly, since the *a posteriori* probability of a hypothesis is a relative frequency, it seems quite legitimate to insist that acceptable hypotheses in science have at least as high a relative frequency as do any of those accepted in the past. Nevertheless, I want to emphasize that success in terms of *a priori* ontological plausibility, and success in terms of ascertained *a posteriori* probability both require methodologies and technologies that "work": in this case, work in the sense of providing epistemologically viable truth-value determinations. This means that the prior probability of a hypothesis is given by the relative frequency of the success of *the ways in which it is investigated*. The success of science is the success of scientific methodologies in solving problems, no matter what end is served by any particular solution. If we suppose that, whatever else might have been the aboriginal intent of science, science means to tell us the truth about the world, to explain the world, to give us trustworthy ways of moving about in the world, then it is to the success of science as a way of solving our problems, *and* as an instrumentality in the orientation of our lives, to which we must turn our attention. In this very specific way, epistemology also yields to methodology, and questions of truth turn into questions of success in practice, including, quite basically, among others, moral questions.

These provisional comments on a model of science as an institutionalized form of inquiry are meant to reveal to both historians and philosophers of science that their special interests are all of them fully justified by the character of institutionalized science, and that the fraternal rules of work in the two disciplines fully justify judgements of success or failure made entirely independently of any questions of satisfying aims larger than those circumscribed by the limits of each discipline. But beyond the special interests of those working within the two disciplines, lies an enriched context of the methodological and epistemological dimensions of science that are free of institutional control, and hence require us to take quite seriously questions of systematic intent; questions of aims and purposes. So regarded, science as a methodology becomes indistinguishable from its pursuit of this or that goal, and, paradoxically, the instrumentalities of science, its technologies, will again be seen to play a central role. These deinstitutionalized features of science, in other words, are only available to be seen when science as one kind of institution is seen as being in competition with other institutions. Perhaps, it is no longer socially realistic to ask whether scientific work in institutionalized form should continue to exist, but it is certainly both reasonable and socially required that we ask how much science there should be, and beyond what point we should continue to support the institution

In the pursuit of answers to such questions, the internal demands of institutionalized fraternal rules of procedure will not be satisfactory except to scientists and others who seek satisfaction of curiosity. For what is now involved is the examination of the credentials of science as a legal individual operating in a social contest against other institutions as legal individuals. This examination will have to lay stress upon the intimate connection between institutionalized science and technology, a connection so deep as to amount in some cases to an interdependency, if not a complete identification. The examination will also reveal a new social role, that of *scientific politician*. Science will need to be defended, but only members of the institution are fit to undertake the defense – only those who have mastered application of the fraternal internal rules will be able to defend the faith now that the institution is under social scrutiny.

What I am suggesting does most emphatically *not* invite replacement of philosophy of science and history of science by sociology of science. Those who apply my model will need to learn the social facts about science, but no replacement of one legitimate field of study by another is either required or envisaged. What I have been arguing for is a twofold thesis of the internal institutionalized priority of seeking satisfaction of curiosity in science and other disciplines on the one hand, and the continuity of methodology and epistemology, as defining properties of the functional activity of science, *viewed outside* the institution, on the other. Historians and philosophers, although they may not share much *in detail* in their work inside their institutions, nevertheless share a great deal when their own activities, as well as those of the scientists, are seen as operating *outside* the given disciplines, and as competing with other forms of social activity, in any way at all.

Appraisals of the worth of a scientific product (a solved problem or a discovered "truth") can thus be undertaken in two fundamental ways. Appraisals of the internal success of a scientific result largely have reference to the fraternal rules of operation typical of the discipline. Although the rules of procedure are mainly conservative, a certain freedom of movement – realization of the urge to introduce novelty – is usually permitted. But what counts is that the result "solves the problem" generated by one of the prevailing research programmes in operation within the discipline.[8] Appraisals of the external worth of the scientific product always reintroduce expected utilities of the research, thus bringing into play various moral and other social cautions. At this level of appraisal, the technological

[8] This is the prevailing theme of Laudan's *Progress and its Problems*.

dependency of science often carries a good deal of the burden of judgement.[9] Finally, the two kinds of appraisals direct attention to the two kinds of scientific success: *success within the discipline* (as measured against the demands of the prevailing fraternal methodologies, preferred ontological commitments, and the like); and *success of the discipline* (as determined by the effectiveness of the institutionalized science in providing solutions to other-than-scientific problems found to be troublesome in the social/political network of institutions operating at a given time).

The short form of my story is this. Both the historians of science/philosophy and the philosophers of science/philosophy have much to offer in any attempt to realize the benefits of my model. Historians can provide us with illuminating accounts of the "data" of appraisal; philosophers can provide us with analyses of the structure of appraisal, the "logics" of method and evaluation. Indeed, there is no reason why – in the best cases – the two tasks cannot be undertaken jointly, often even by the same researcher. But if the historian cannot, among other sciences, study philosophy, there is reason to doubt the reliability of his reading of the history of science/philosophy. Similarly, if the philosopher cannot do conceptual justice to the roots of his enterprise, there is reason to suspect that what appears arid and overly antiseptic in his work is just that – arid and antiseptic. All the rest is, for the most part, simply professional jealousy.

---

[9] For a detailed and enlightened discussion of the idea of the technological dependency of science see Rescher's *Scientific Progress*, 132-192.

CHAPTER 12

SCIENCES AND PSEUDOSCIENCES: AN ATTEMPT AT A NEW FORM OF DEMARCATION[1]

The Demise of the Demarcation Problem

In an article appearing in an earlier Festschrift honoring Adolf Grünbaum (Laudan 1983),[2] Larry Laudan argued persuasively that the attempt to distinguish science from pseudoscience by employment of one or another of the proposed demarcation criteria is now dead. What he had specifically in mind is that demarcation criteria employing verificationist and falsificationist theories of meaning have failed to accomplish the required goal.[3] However, he obviously

---

1 [Previously published in John Earman, Allen I. Janis, Gerald J. Massey, and Nicholas Rescher (eds.) *Philosophical Problems of the Internal and External Worlds: Essays on the Philosophy of Adolf Grünbaum* (Pittsburgh: University of Pittsburgh Press, 1994), pp. 163-185]

I mention in the text some of those who have helped to shape my thinking about science. The contribution of Adolf Grünbaum is simply incomparable, and is here acknowledged with warm thanks. I have also learned about the science-pseudoscience divide from my conversation and correspondence with Andrew Lugg (see Andrew Lugg, "Bunkham, Flim-flam and Quackery", *Dialectica* 41 (1987): 221-30, and "Stuff and Nonsense: Pseudoscience as Madness", forthcoming).

2 Larry Laudan, "The Demise of the Demarcation Problem", in R. S. Cohen and L. Laudan (eds.), *Physics, Philosophy and Psychoanalysis: Essays in Honor of Adolf Grünbaum* (Dordrecht: Reidel, 1983), pp. 111-127.

3 The employment of theories of meaning to help in drawing the line between science and nonscience now seems a lost cause. The effort was to show that the formal and empirical sciences contain sentences that are cognitively significant, whereas the sentences of nonsciences are in some respect non*sense*. However, "sense" is a notoriously ambiguous word. The attempts at demarcation along such lines (whose ancestor is the Kantian strategy to be discussed in the next section) could never answer the rejoinder of Nietzsche:

> It is no different with the faith with which so many materialistic scientists rest content nowadays, the faith in a world that is supposed to have its equivalent and its measure in human thought and human variations – a 'world of truth' that can be mastered completely and forever with the aid of our square little reason. What? Do we really want to permit existence to be degraded for us like this – reduced to a mere exercise for a calculator and an indoor diversion for

retained confidence in what he called "our intuitive distinction between the scientific and the non-scientific" (ibid., 124) and admitted that some future successful attempt to mark the difference could not be ruled out. The intuition referred to provides a rich catalog of pseudosciences, "flat Earthers, biblical creationists, proponents of laetrile or orgone boxes, Uri Geller devotees, Bermuda Triangulators, circle squarers, Lysenkoists, charioteers of the gods, *perpetuum mobile* builders, Big Foot searchers, Loch Nessians, faith healers, polywater dabblers, Rosicrucians, the-world-is-about-to-enders, primal screamers, water diviners, magicians, and astrologers" (ibid., 121). Evidently, these assorted pseudosciences make "crank claim[s] which make ascertainably false assertions"(ibid.). One might reasonably add: *if literally construed*.[4]

The concept of literal construal of claims to knowledge will play an important role in my argument for a new kind of demarcation criterion. But first, I want to review some ways of distinguishing between science and pseudoscience that do not in any obvious ways invite exhumation of the corpses Laudan has rightly buried. What continues to nag at me is Laudan's adoption of a centuries-old intuition. Many of us have this same intuition. I am convinced that in some way we can express its message in fairly precise terms. Although we may regard as inadequate certain forms of verificationist and falsificationist *criteria* of demarcation, the *problem* of demarcation seems to remain.

I will briefly review four philosophical strategies that have some merit as attempts to distinguish science from nonscience. All four of the strategies have a bearing on the new strategy I will propose.

> mathematicians? Above all, one should not wish to divest existence of its *rich ambiguity* (*seines vieldeutigen Characters*): that is a dictate of good taste, gentlemen, the taste of reverence for everything that lies beyond your horizon. That the only justifiable interpretation of the world should be one in which *you* are justified becauise one can continue to work and to research scientifically in *your* sense (you really mean, mechanistically?) – an interpretation that permits counting, calculating, weighing, seeing, and touching, and nothing more – that is a crudity and naiveté, assuming that it is not a mental illness, an idiocy. Would it not be rather probable that, conversely, precisely the most superficial and external aspect of existence – what is most apparent, its skin and sensualization – would be grasped first – and might even be the only thing that allowed itself to be grasped? A 'scientific' interpretation of the world, as you understand it, might therefore still be one of the *most stupid* of all possible interpretations of the world, meaning that it would be one of the poorest in meaning. (F. Nietzsche, *The Gay Science*, translated by Walter Kaufman (New York: Vintage, 1974), sec. 373)

4 Over the centuries, the list has not changed that much. Here is one from Reginald Scot, *The Discovery of Witchcraft* (London: n.p., 1665), "the Knavery of *Juglers*, *Conjurers*, *Charmers*, *Soothsayers*, *Figure-Casters* [astrologers], *Dreamers*, *Alchymists* and *Philterers* [sellers of erotic love potions] ..." (p. i).

Forms of Reliable Knowing, Science, and Mental Aberration: The Kantian Strategy

Kant's well-known critique of classical substantive metaphysics employs an explicit form of demarcation between science and nonscience of the following kind.[5] First, Kant's analysis of human cognition seeks to locate those species-universal mental structures that define reliable knowing. These structures are the now familiar forms of space and time and the categories functioning as a priori, hence necessary, forms of the possibility of knowledge. Kant argues that this complex cognitive machinery, applying as it does only to items given in sensation, effectively distinguishes between reliable knowledge (compliant with the cognitive structures) and pseudoknowledge (such as that alleged to be knowable in classical metaphysics). Furthermore, it was Kant's view that the premier instantiation of the formalism of spacetime and the categories is given by Newtonian physics: *The* most reliable form of human knowing is mathematical physics.[6]

At various stages in his career, acceptance of this conclusion meant for Kant that biology can only be construed as a research strategy pursued in the interests of obtaining good taxomony, not a full science (the position of *Critique of Judgment*), and that chemistry is not a science but an artful classification scheme (the position of *Critique of Pure Reason* and *Metaphysical Foundations of Natural Science*).[7] It also meant that substantive metaphysics has no reliable epistemic credentials, and that some claims to knowledge of that which lies beyond experience are indistinguishable from claims characteristic of mentally deranged persons (Kant's judgment on the claims of Swedenborg to possession of ways of access to a world of spirits). It was also part of Kant's strategy to propose that various forms of mental activity that deviate from, ignore, or distort application of the epistemic formalism can be classified as maladaptive mental disorders. Any such forms of maladaptive behavior will produce claims that cannot be justified as scientific (see Butts, *Kant and the Double Government Methodology*, 282-318).

In summary, here is Kant's strategy: identify the norms of reliable knowing, show that these norms are instantiated by the best science of the day, then classify pseudosciences as aberrations of these norms (metaphysics and other claims to transcendent knowledge), which in some cases involve genuine mental

---

5 For a more detailed account see R. E. Butts, *Kant and the Double Government Methodology: Supersensibility and Method in Kant's Philosophy of Science* 2nd ed. (Dordrecht: Reidel, 1986).

6 See M. Friedman, "The Metaphysical Foundations of Newtonian Science", in R. E. Butts (ed.), *Kant's Philosophy of Physical Science: Metaphysiche Anfangsgründe der Naturwissenschaft, 1786-1986* (Dordrecht: Reidel, 1986) pp. 25-60, and R. E. Butts, "Introduction: Kant's Quest for a Method for Metaphysics", pp. 1-24, and "The Methodological Structure of Kant's Metaphysics of Science", pp. 163-199, in *Kant's Philosophy of Physical Science*.

7 In the *Opus postumum* Kant altered this assessment of the cognitive status of chemistry, and tried to give a justification of scientific status of the work of Lavoisier.

illness. There is much to commend Kant's strategy, but unfortunately in its Kantian form it cannot be sustained. Surely we cannot require that the norms of reliable knowing must be instantiated by any science of a particulr period if what we take science to be is the *content* of a particular successful theory. This defect has led many to look for distinguishing marks of science in its *methods* and what have come to be called its *virtues*. Before turning attention to such strategies, it must also be remarked that the Kantian form of demarcation can only be applied with profit after cognitive science and psychopathology have joined forces. In the end, the Kantian strategy cannot be successfully accomplished philosophically; it needs to be itself a part of science. This consequence should be applauded, for it cautions us to look for demarcation and the definition of cognitive normalcy, if they are to be found at all, not in a priori philosophical arguments, but in the results of combined theoretical and empirical investigation. The failure of the Kantian program as a philosophical account of demarcation itself points the way to some empirically ascertainable distinguishing marks of the sciences.

Prediction and Control: The Rescher Strategy

Many philosophers and scientists have directed our attention away from the specific content of sciences, and toward the expressed goals of scientific inquiry. The goals are many: accuracy of description, mathematical precision in the formulation of theory, retrodiction, and taxonomical adequacy. Two goals, however, are thought to occupy prominent positions largely unique to science: accuracy in prediction, and success in control of the environment. Nicholas Rescher presents a particularly clear and persuasive argument to the conclusion that what he calls "explanatory frameworks",[8] including the frameworks of the sciences as well as, for example, numerology, astrology, and black magic, can each be criticized *en bloc*, but only a posteriori. The sciences win in the competition, not because of superior content, but because of pragmatic success in prediction and control that is unmatched by other explanatory frameworks.[9]

Part of Rescher's strategy is clearly correct. We have learned that metaphysics (and Kant's transcendental arguments) provide no a priori grounds for favoring certain sciences over alleged nonsciences, or, alternatively stated, for preferring a definition of cognitive normalcy over claims to the possibility of heightened modes of awareness, or altered states of consciousness. Furthermore, the failed empiricist effort to locate a demarcation criterion that would single out science as the sole dispenser of cognitively significant sentences further confirms the absence of a priori grounds that would allow a decision between explanatory

---

[8] Nicholas Rescher, *Scientific Explanation* (New York: The Free Press, 1970), pp. 135-45.

[9] See Nicholas Rescher, *Methodological Pragmatism: A Systems-theoretic Approach to the Theory of Knowledge* (New York: New York University Press, 1977).

frameworks.

It seems unlikely, however, that one can dismiss astrology, appeals to narrative discourse, or fundamentalist creationism, or psychoanalysis, solely because these explanatory frameworks fail as accurate systems of prediction leading to at least partial control of aspects of the environment. It all depends on what is at stake. Explanatory frameworks operate to further chosen objectives. If we know what these objectives are, we can show, by admittedly a posteriori means, that those ends cannot be realized in ways promised by the given explanatory framework. Pseudosciences are shown to be defective in explanatory potency by discovering that they possess impoverished means for realizing *aims they themselves articulate*.

In different ways, the failure of the Kantian strategy and the partial success of the Rescher strategy point in the same direction. We must look for demarcation by paying attention to what kinds of things conceptual frameworks are attemting to accomplish by saying what they do.

Science and "Cognitive Accountability": The Grünbaum Strategy

Adolf Grünbaum's many writings on the epistemological status of psychoanalysis[10] offer no new theory of demarcation. However, in other papers, his work does seek to illuminate the kinds of problems that must be faced by those who would deal with the demarcation problem. In several papers,[11] he has argued vigorously against Popper's claim that psychoanalysis is a pseudoscience because it is not falsifiable. Popper's conclusion is of course based on his acceptance of the falsificationist criterion of demarcation. What Grünbaum shows is that a thoroughgoing *inductivist* study of psychoanalysis reveals that some of its hypotheses are indeed falsifiable, and falsified, thus refuting Popper's claim that Freud's theories are not falsifiable, and at the same time casting doubts on the merits of the falsificationist criterion of demarcation.

Grünbaum's study[12] of the hermeneuticians' attempts to free Freud's

---

10 See A. Grünbaum, "Epistemological Liabilities of the Clinical Appraisal of Psychoanalytic Theory", *Noûs* 14 (1980): 307-85; "Logical Foundations of Psychoanalytic Theory", *Erkenntnis* 19 (1983): 109-52; *The Foundations of Psychoanalysis: A Philosophical Critique* (Berkeley and Los Angeles: The University of California Press, 1984).

11 See for example "The Degeneration of Popper's Theory of Demarcation", *Epistemologica* 12 (1989): 235-60

12 "Why Thematic Kinships between Events Do Not Attest Their Causal Linkage", in J. R. Brown and J. Mittelstrass (eds.), *An Intimate Relation: Studies in History and Philosophy of Science Presented to Robert E. Butts on His 60th Birthday* (Dordrecht: Kluwer, 1989), pp. 477-94; and "'Meaning' Connections and Causal Connections in the Human Sciences: The Poverty of Hermeneutic Philosophy", *Journal of the American Psychoanalytic Association* 38 (1990): 559-77.

theories of what Freud himself insisted is the *natural* scientific status of those theories also has an indirect bearing on the problem of demarcation. Jaspers, Habermas and Ricoeur think of the clinical situation in psychoanalysis as one eventuating in hermeneutical insight into kinships of meaning, where such affinities are also regarded as *causal*. In this way, the hermeneuticians seek to free Freud's theories from the corpus of natural sciences so that they can take their place among the human sciences. Grünbaum demonstrates, again by employment of inductivist methods of analysis of causal connection, that connections between meanings, even when very strong, do not suffice of themselves to establish causal connections. Thus the hermeneuticians' efforts to show that psychoanalysis is a pseudo *natural* science are unavailing. Such efforts, in effect, simply reinstate long-standing epistemological problems associated with the human sciences, for example, epistemic claims on behalf of empathy or *Verstehen*.

Grünbaum's contribution to the solution of the demarcation problem consists in his invitation to learn just how difficult it is to establish the scientific credentials of a theory. He does this, not by offering a criterion of demarcation, but by demonstrating that psychoanalysis, however brilliantly conceived, is *inadequate* science. His arguments are designed to show, not that psychoanalysis is a pseudoscience, but that the scientific credentials of the theory are deeply flawed by mistakes in logic, epistemological and ontological sleights of hand, and failures to credit disconfirmations when they occurred. Grünbaum refers to "the sort of cognitive accountability that is featured by the natural sciences"[13] (and which the hermeneuticians seek to bypass by freeing the human sciences of the constraints normally imposed on research in the natural sciences).

The Grünbaum strategy thus adheres to my suggestion that in examining any theory (scientific or pseudoscientific), the articulated aims of the theorizing must be taken seriously into account. What are the articulated aims of Freudian science (for Grünbaum, *all* science)? What is the cognitive accountability featured by the natural sciences? The answer will surely go something like this: A theory in a natural science should have testable consequences, because, before all else, an empirical theory must be confirmed or disconfirmed by observations; a theory should be logically rigorous; a theory should be simple. There are other such "virtues" of theories in the natural sciences (Hempel[14] calls them "*desiderata*"). It may also be true that not every theory accepted by natural scientists possesses all of the virtues. How many the theory must possess, and how basically involved they are in a given theory, are perhaps variable. In general, however, cognitive accountability as it is envisaged by Grünbaum involves essentially the accessibility of theories to rigorous and detailed criticism based on canons of inductive and

---

[13] "'Meaning' Connections and Causal Connections in the Human Sciences: The Poverty of Hermeneutic Philosophy", *Journal of the American Psychoanalytic Association* 38 (1990), p. 562.

[14] C. G. Hempel, "Valuation and Objectivity in Science", in R. S. Cohen and L. Laudan (eds.), *Physics, Philosophy and Psychoanalysis: Essays in Honor of Adolf Grünbaum* (Dordrecht: Reidel, 1983), pp. 73-100.

deductive logic.

Marks of a Good Hypothesis: The Hempel Strategy

Hempel lists *desiderata* of scientific hypotheses, suggesting that they are imprecise constraints on scientific theory choice, "Among them are the following: a theory should yield precise, preferably quantitative, predictions; it should be accurate in the sense that testable consequences derivable from it should be in good agreement with the results of experimental tests; it should be consistent both internally and with currently accepted theories in neighboring fields; it should have broad scope; it should predict phenomena that are novel in the sense of not having been known or taken into account when the theory was formulated; it should be simple; it should be fruitful."[15]

Hempel concedes that these desiderata are vague, and mentions that in earlier writings he had argued that they are arational (not irrational) constraints on theory choice. However, he thinks that his analysis should lead us to conclude that there is something left over that can count as objective traits of competing theories, for example, "the considerable role that precise and rule-governed reasoning does play in the critical appraisal of competing theories, which requires among other things a rigorous derivation of experimental implications and the performance of experimental tests that have to meet appropriate standards But even the considerations adduced in appraising the satisfaction of the vaguer desiderata are typically perceived, it seems to me, as expressing not just individual taste or preference, but objective, if only roughly specified, traits of competing theories" (ibid., 95).

Thus Hempel agrees with Grünbaum that there are standards of cognitive accountability, and provides an outline of how these standards apply. In doing so, he seems to me to be providing grounds for distinguishing between sciences and those nonsciences appearing on Laudan's list. Apart from the desiderata, Hempel perceives in science some "quite generally acknowledged norms" (ibid., 93), for example, the requirement that scientific reasoning must comply with standards of deductive logic and the interdiction against logical inconsistencies; norms governing methods of measurement and the testing of statistical hypotheses. Such relatively precise norms, together with the desiderata, in effect define scientific appraisal of theories.

Bas van Fraassen[16] chooses to call the Hempel desiderata "virtues" of theories, and argues that the virtues have a pragmatic, not an epistemic, status. Consistency, empirical adequacy, and empirical strength are designated as epistemic traits of scientific theories. But, argues van Fraassen, in accepting a

---

15 C. G. Hempel, "Valuation and Objectivity in Science", op. cit., pp. 87-88.

16 B. C. van Fraassen, *The Scientific Image* (Oxford: Clarendon, 1980), pp. 87-89.

theory, we are not just believing it, we are also committing ourselves to certain wagers: that the research program that generated the theory is worth pursuing, that all relevant phenomena will be captured by the theory, and the like. Epistemic features of a theory (Hempel's "norms") have to do with a theory's relationship to the world; pragmatic virtues of a theory have to do with what we as human knowers want the theory to achieve over and above our belief that it is empirically adequate.

What We Learn from the Four Strategies

A bit of summing up is in order. The strategy of Kant, although it may be the most desirable one, *if achievable*, must wait upon further developments in the sciences before it can be sustained. As a philosophical program it cannot establish what it sets out to conclude: that there is just one set of species-universal conditions on the basis of which we can arrive at reliable knowing, defined by methodologies appropriate to a science regarded as fixed and finished.

As I have suggested, the strategy of Rescher has some value. It invites us to direct attention away from the content of sciences and their competitors and to features of method. In addition, this strategy, as we have seen, replaces a priori demarcation criteria with a posteriori evaluations of success. However, we are now in a position to see that accuracy in prediction and control of the the environment are desiderata or virtues of theories, and must be taken into account in the full appraisal of competing theories, but that they cannot by themselves distinguish between sciences and nonsciences. If we leave aside non-epistemic goals scientific or nonscientific theories might serve – for example, improvement of various aspects of human and other animal life – and direct our attention solely to epistemic goals (those whose realization employs, in the sciences, the norms Hempel mentions), and construe these by reference to empirical adequacy, empirical support, and consistency, accuracy in prediction and success in control are clearly not decisive factors in marking off science from nonscience.

What the strategies of Grünbaum and Hempel (and of van Fraassen) enjoin is attention to the norms of cognitive appraisal that must be actually operative in determining a theory's adequacy as a set of claims *about the world*. Here, the constraints on theory are the epistemic ones of deductive consistency (and of compliance with norms of nonfallacious ordinary reasoning) and empirical *support*. Therefore, for purposes of fair comparison of theories thought to be scientific with those thought not to be scientific, we need to emphasize that, other nonepistemic things being equal, the aim of all such theories is *to tell the truth about some features of the world*. What is to be investigated, then, are the credentials a given theory has offered on behalf of its entitlement to be taken seriously as a set of empirical claims. In other words, we need to know what a given theory can offer in the way of support for its claims to empirical knowledge. In what follows, I will show a way of making a fairly precise distinction between those theories with acceptable credentials, and those lacking them.

A New Strategy of Demarcation

> MAN is the Interpreter of Nature, Science the right interpretation - William Whewell, *Novum Organon Renovation*

Man interprets nature in many ways, by means of science, poetry, walks in the woods, sailing on dangerous seas, telling age-old stories, seeking spirits in trees and clouds, suns and pestilences. I am interested here only in those interpretations that are cognitive claims, or texts. Texts in this sense are of course also subject to interpretation. What is to be avoided is thinking of nature as a text. Nature is interpreted, but is not a text. Symptoms, microscope slides containing slices of tissue, animal footprints, all require interpretation, but are not texts.[17] If one thinks of all the kinds of texts there are, the problem of learning how to interpret them arises: Given a text of one sort or another, what are the permissible interpretations? Goodman has of course raised this question in a number of his writings.[18] I think we should begin to take it seriously.

Goodman seems to me correct in thinking that texts are defined syntactically, and that interpretation of meaning is a matter of semantics. Texts are normally strings of inscriptions or vocal sounds. There are exceptions. Some texts are pictorial, and only context will distinguish text from design. What the old grammarians used to call *hypotyposis* or *subjectio sub adspectum*, referring to representational exhibition of meaning with sensuous clarity, encompassed both metaphor and pictorial image, both poetic symbol and visible sign. Thus I will want to include mandalas and international road signs as texts.[19] We can then generate

---

[17] Anything that can be counted as a system of symbols subject to interpretation is a text. this list will be very long, but will surely contain the following: proofs of theorems, reports of results of observation or experiment, narratives (histories, news releases, novels, travelogues, stories told in psychoanalytic clinical situations), police reports, lawyers' arguments, judges' decisions, poems, clinical observations, mandalas, cryptograms, messages from another world (as in seances, contact with extraterrestrials), pictograms (cave and rock paintings), works of visual art, plays, operas, musical scores, astrological charts and advice, records of divine revelations, ceremonies (Christmas, the Festival of Lights), tribal lore transmitted verbally, the writings of philosophers, metal coins, political speeches, sermons, academic lectures, street and highway signs.

[18] See N. Goodman and C. Z. Elgin, *Reconceptions in Philosophy and Other Arts and Sciences* (Indianapolis: Hackett, 1988), part 2, section 3.

[19] Although again context is important. In the neighborhood of my birthplace in Lancaster County, Pennsylvania many Amish towns have charming names. The signs read "Paradise," "Intercourse," "Blue Ball," "Bird-in-Hand." Undergraduates at the University of Pennsylvania in Philadelphia are fond of stealing these signs for use as decorations in their dormitory rooms. What were

a long list of different kinds of texts, and we can begin to wonder how many good or right interpretations are possible for each kind. A sentence describing chemical and morphological features of amethyst samples appears to be correctly interpretable in only one way, or in very few permissible ways, whereas Shakespeare's *Hamlet* may have number of right interpretations. The Hamlet of Olivier and the Hamlet of Zeffirelli are both acceptable readings of Shakespeare's text.

These considerations lead us to the idea of *a range of permissible interpretations* of a given text, and to the equally suggestive idea that kinds of texts can be distinguished from one another by *the size of that range*. Think then of a scale of texts ranging from those for which only one interpretation is permissible to those for which any number of interpretations are permissible. At one end of that scale will be, perhaps among others, those texts stating claims in formal systems of mathematics and logic, where the texts have only one permissible interpretation, just because no semantics is involved. To have only one permissible interpretation is equivalent, I think, to requiring no interpretation at all. International road signs will be at this extreme end of the scale, as will texts stating standardized rules of a game, say, chess.[20] At the other extreme end of the scale. what one might call the limit of interpretation, is nonsense or absurdity, that textual gibberish for which any interpretation at all will do. I take it that allowing any interpretation at all is equivalent to having no permissible interpretations. So we might say that gibberish is "beyond interpretation". At other places along the scale will be all of the other kinds of texts, each with a different range of permissible interpretations. Our question now becomes as follows: Where on this scale do we place the various texts of the natural sciences, and where, the texts of the various enterprises thought intuitively to be classed as pseudosciences? Since it is clearly in the interests of all parties to this dispute to have their range of textual interpretations as close to 1 as possible, a collateral question is as follows: How, in the cases of those texts where truth-telling is a central goal, does one limit the range of required hermeneutic insight, the range of permissible interpretations? We will see that the answers to both questions relate also to what we can reasonably mean by support for textual claims of various sorts.

We need to limit ourselves to those texts seeking to tell the truth about the world. They might seek to do other things as well, but for the purposes of the present exercise employment of rhetoric ingredient in a given text will be ignored. We are interested, then, only in those texts which, literally construed, can be either

---

signs thus become trophies, and are not to be taken literally.

[20] I am aware that one can be deceived by some road signs, especially those marking miles or kilometers. One summer, a student of mine worked for the Department of Highways; he took delight in placing mileage markers at random, rather that at precisely determined intervals. Max Black used to say that "it isn't a language unless you can lie in it." It behooves all of us to face all texts with a measure of skepticism. Even mathematical texts can be misprinted. No important epistemological point turns on these admissions.

true or false. Notice that I say "*can be*" either true or false. I think it is totally counterproductive to argue for a distinction between sciences and pseudosciences on the basis of acceptance of the form of scientific realism which teaches that to accept a scientific theory is to believe that it is true. The pseudoscientist is always in a position to counter this view with the correct observation that no texts stating views about the world have ever been conclusively established as true, hence none are ever believable because true. This is the thin end of the wedge actually employed by many who challenge the epistemic credentials of the sciences: Since no view has ever been conclusively established as true, mine is as good as yours, and on altogether proper epistemic grounds. Because the pseudoscientist is always in a position to exploit the logical contingency of generalizations over items of experience, van Fraassen's antirealism seems to be the best philosophical standpoint from which to argue with the pseudoscientists. Van Fraassen holds that although the language of science must be literally construed, and hence that such language can be either true or false, acceptance of a theory only amounts to believing the theory is empirically adequate, and possesses certain virtues.[21]

In thus suggesting that van Fraassen's antirealism is a better foil to pseudoscience than is realism, I am not suggesting that this constitutes a telling reason for accepting van Fraassen's constructive empiricism. What I want to emphasize is that if the difference between science and nonscience is to be made out fairly, those of us who accept the intuitive difference had better approach the problems with a very parsimonious ontology and philosophy of theory acceptance. On one point there seems to be no fundamental disagreement: what Hempel takes to be the core norms of natural science inquiry (and which supply the basis for that cognitive accountability Grünbaum thinks to be definitive of natural science) is entirely compatible with van Fraassen's idea that the central features of science have to do with empirical adequacy, empirical support, and consistency. *At least this much* would be wanted by both realists and antirealists, and constitutes enough for us to go on.

Let us then think of the literal construal of a text as accepting that what it says can be either true or false, and as accepting nothing stronger than that. We need this idea to apply both to scientific texts and to those thought to be pseudoscientific, for claims from both quarters seek at a minimum to be empirically adequate, empirically supported, and consistent. Further, we need to endorse van Fraassen's statement that "to insist on a literal construal of the language of science is to rule out the construal of a theory as a metaphor or simile, or as intelligible only ofter it is 'demythologized' or subjected to some other sort of 'translation' that does not preserve logical form".[22]

It is important to note that the only form of translation allowed of sentences in organic chemistry, astrology, and creation science is translation into

---

21 *The Scientific Image*, pp. 10-12.

22 *The Scientific Image*, p. 11.

another natural language. For purposes of fair comparison, all contenders must state texts in one of the natural languages, which can only be translatable into any of the others.[23] There is no hidden meaning lying behind or beyond literally construed language in the sense required here. All of this granted, where do the various literally construed texts fit on the scale of texts? What is it that gives the texts of the natural sciences and those of the alleged pseudosciences different places?

My thesis is that, all things considered, the texts of the natural sciences are closer to 1 on the scale of texts with ranges of permissible interpretations than are the texts of the pseudosciences. This is to say that the texts of the natural sciences permit only one or very few interpretations, whereas those of the pseudosciences can be fit to many more legitimate interpretations. To see why this is so it is only necessary to return attention to Hempel's norms and Grünbaum's principles of cognitive accountability and to the idea of support for literally construed texts.

Science and Stability of Meaning

Hempel's desiderata and Grünbaum's implicit principles of cognitive accountability are derived mainly from work in the physical sciences, where the literally construed claims are stated with great care, are sometimes quantitative, and can even be very dangerous in consequences if not understood and applied with precision. Such claims, furthermore, require for their support experimental consequences that require exact measurements of distances and times. The language in experimental physics, chemistry and other natural sciences is thus as decisively referential (fixed in meaning) as it is possible to be, and interpretations of that language if they are not to be taken to be completely frivolous, approach 1.[24] All properly formulated scientific discourse has a range of 1, or very near 1. This is so because such discourse must at some point make contact with public, shared, controlled observation, and because of the stability of meaning of its technical

---

[23] I must admit that sentences in scientific reports are more readily translatable into other natural languages partly because much of the terminology is standardized, and the mathematics employed is normally semantically transparent. Such ease in moving from one language to another in science is a stroke of good fortune not also available to the pseudosciences. Translations of texts in astrology, creationism and the like might preserve logical form, but they inevitably introduce ambiguities and nuances not present in the source language. To which version of the Bible should creationists direct us? I will return to this problem later.

[24] A pragmatic consideration here gives weight to Rescher's emphasis upon accurate prediction: If some experiments are not carried out with the greatest attention to all details, the experimenters are at peril. What is lost if an astrological prediction turns out to be false? Perhaps the client may murder the astrologer. That is quite different from the death of a careless experimenter.

terms. This constitutes its support.[25]

We need not go to the highly sophisticated reaches of chemistry and physics, and to their applications in practical affairs, in order to understand the phenomenon of stability of meaning in the sciences.[26] Consider, for example, one of Pasteur's experiments designed to help our understanding of the impossibility of spontaneous generation. Pasteur dissolved a crystalline salt of ammonia, crystalline sugar, and phosphates from the calcination of yeast in pure distilled water, and then sowed into the liquid a few spores of *Penicillium* or some mold.[27] The spores germinated within two or three days, filling the surface of the liquid with flakes of mycelium, where they produced healthy vegetation. Pasteur's report continues, "The precaution of employing an acid salt of ammonia hinders the development of Infusoria, which would soon stop the progress of the little plant through absorbing the oxygen of the air, without which the mould cannot thrive. All the carbon of the plant is obtained from the sugar, which is slowly used up, all its nitrogen from the ammonia, its mineral matters from the phosphates. In this matter of the assimilation of nitrogen and phosphates there is thus a complete analogy between ferments, moulds, and plants of complicated organization."[28]

Pasteur next reports that if any of the dissolved principles are omitted, the growth of the fungi is arrested. The conclusion is that fungi do not obtain carbon in the same way as do phanerogamic plants, "They do not decompose carbonic

---

[25] It might be argued that stability of meaning of technical terms is assured by the *community* of scientists, and that similar communities of astrologers and creationists and others create their own assured stability for the meaning of their technical terms. I would not deny that fraternal conditions play some role in fixing meaning. What one needs to consider here is the nature of the conditions of entry into different fraternal groups. Rigorous education and satisfaction of high intellectual standards form part of those conditions for scientists. No such training and accomplishment are required of young astrologers; all that is needed for entry into the fundamentalist community is an unwavering faith in the literal correctness of biblical claims. Indeed, further study of such conditions of entry into intellectual and quasi-intellectual communities might highlight still more ways in which the sciences and the pseudosciences differ fundamentally.

[26] I am emphasizing features of the discourse of experimental and observational natural science, because I am dealing only with those texts that seek to tell truths about the world. Such an emphasis, however, should not disguise the fact that within *theoretical* natural science, interpretations of theoretical constructs are themselves part of the internal dialectic of physical science. Thus, for example, the concept of gravity in classical Newtonian physics can be interpreted as action at a distance, or as involving electromagnetic fields. So far as the mathematics and the empirical support of the classical theory are concerned, these different interpretations are equivalent, and play only heuristic roles. Like Kant's ideas of reason, they have no empirical truth-telling content. I am grateful to my colleague Robert DiSalle for calling these matters to my attention.

[27] The liquid consisted of 100 parts pure water, 10 parts sugar, 0.2 to 0.5 of ammonia, and 0.1 of ash of yeast.

[28] M. L. Pasteur, "Memoir on the Organized Corpuscles which Exist in the Atmosphere", in W. C. Dampier and M. Dampier (eds.), *Readings in the Literature of Science* (New York: Harper & Brothers, 1959), pp. 217-29.

acid; they do not set free oxygen. The absorption of oxygen and the setting free of carbonic acid; are on the contrary the necessary and permanent activities of their life. these facts give us precise ideas on the mode of nutrition of fungi ..." (ibid.). This is precise enough to enable brewers, wine and vinegar makers and diary farmers to introduce controls on their procedures based on scientific knowledge. But that is only part of the point. the larger part is that it is not possible to understand the elements of this simple inductive experiment on the basis of several interpretations. Too much precision and exact knowledge exists at every point.

What is being investigated depends upon well-understood aspects of chemistry and biology: the nature of distilled water, of calcinated yeast, of sugar, of tartrate of ammonia, the required quantities of these chemicals, the classification of plants with organs of reproduction as phanerogamic, the behavior of Infusoria, and the kind of nutrition involved in such plants. all such matters are expressed in language whose meaning is fixed; there is no room for gratuitous interpretation. There is room for disagreement, but that is another matter. Notice also that I have deliberately employed a translation of Pasteur's paper, first published in *Annales de Chemie et de Physique* in 1862. The translation does not "demythologize," it simply transfers exact language from one natural language into another. For purposes of exact science, all natural languages are one language, construed literally.

One final point about the experiment is that what Pasteur wrote can be *misunderstood*. Furthermore, Pasteur could have set out to deliberately *deceive* us. I see no insurmountable problems here. With respect to texts whose range of interpretations approaches 1, misunderstanding can be corrected, perhaps not easily, but surely in time. A misunderstanding is, in the end, a wrong interpretation, and wrong interpretations are not permissible interpretations. Any text can set out to deceive us, even in the formal sciences. Here again, the detection of deception, although in some cases it may be very difficult to realize, is always in principle realizable for those texts whose range of interpretation approaches 1. The possibility of fraud does arise in the work of natural scientists; it can never arise in the work of poets.[29]

What about astrology, creationism, and other alleged pseudosciences, can their texts be literally construed and empirically supported, establishing that the ranges of interpretation are close to 1?

Astrology and Its Support

[29] However, notice David Hume's amazing attempt to rescue some literal sense for poetry, "Poets themselves, tho' liars by profession, always endeavor to give an air of truth to their fictions; and where that is totally neglected, their performances, however ingenious, will never be able to afford much pleasure" (*A Treatise of Human Nature*, edited by L. A. Selby-Bigge, Oxford: Clarendon Press, 1951, p. 121).

Astrological texts are charts based on calculations of positions of heavenly bodies at various times, which relate such positions causally to events in people's lives. These results are often stated in sentences that assess trends. One can find such sentences in most newspapers published in the world today. Some condemn astrology because it does not accurately predict precisely events that will determine the course of someone's life. But careful astrologers make more cautious claims, "Astrologers are not magicians. They do not deal in specific events and must always take care that no client is injured through taking their advice too literally"; and:

> Serious astrologers do not "predict" events; they assess trends. When mentioning potential hazards or beneficial periods, the astrologer must always remember to put these matters carefully to his client. He should never state: "You will have an accident on the twenty-first of the month," but he can say that: "On the twenty-first of the month you are likely to be somewhat prone to risks, so take additional care when handling hot dishes or sharp knives." Of course, in this way astrologers do lay themselves open to the criticism of using generalizations. It must be made clear to a client that progressions[30] are rather like weather forecasts, and should be used in much the same way.[31]

A host of questions arise. If astrological results cannot predict, how do we determine the success of astrological claims? If the advice given by astrologers is general, how can it avoid vagueness and loss of clarity? If the advice does not speak to specific events, are we not invited to interpret it in very many ways – that is, to what I take it to mean for my life, to what you take it to mean for yours, and so on? Are not publishing the results of theoretical and experimental work and advising clients two entirely and *importantly different* matters? Does the casting of charts have any valid relationship to the advice given a client? What, after all, is the relationship between the calculations and the general advice? The advice could, for all we know, eventuate from many other kinds of sources. Is it really possible, given these questions, to construe astrological advice literally?

Consider the following item from *The Globe and Mail* (1991):

> *Having a Hussein Week* Saddam Hussein's birthday is April 28. Here is some of Jeane Dixon's advice to Taurus last week:
> Jan. 15 (UN's Tuesday deadline): Indulging your whims could prove costly today. Think about the image you want to present.
> Jan. 16 (U.S. planes hit Baghdad, including its communications centre): Spending time alone will help you extricate yourself from a difficult situation.
> Jan. 17 (Iraqi missiles strike Israel): Do what must be done at home. You can easily catch up with neglected tasks.

---

30 Birth charts are "progressed" by, for example, taking the positions of the planets on the first day after birth to relate to the conditions in the first year after birth, the positions in the fifth day after birth, to the fifth year after birth, and so on for all days after the birth date. These progressions are employed to assess trends in the life of a person.

31 D. Parker and J. Parker, *The Compleat Astrologer* (New York: Bantam Books, 1979), p. 154.

> Jan. 18 (Jordanian parliament brands United States the Great Satan): A close relationship intensifies; two minds think as one. You sense what will happen next. Now is the time to make important connections.

I suppose *The Globe and Mail* thought this piece to be humorous; if each bit of advice is construed literally, then it would seem that Saddam has failed to heed the warnings. However, even if construed literally, each piece of advice is permissibly subject to different interpretations since the advice is general and does not speak to Saddam or any other individual. (Another question is if the advice is this general, over what instances does it generalize? the life charts of everyone born under the sign of Taurus? the character traits mythologically associated with each sign of the Zodiac? The charts and the traits are themselves open to endless interpretation.) Saddam need not have thought that failure to meet the UN deadline was whimsical; he might have wanted to project exactly the image of one who acts in defiance of the deadline. Perhaps he thought that remaining in his bunker was exactly the way to prevent his own destruction. Perhaps he thought that sending missiles to Israel was exactly the way to look after the home front.

How, then, does the astrologer attempt to limit the range of permissible interpretations of the claims involved in the advice? One way is to insist that the kind of advice given in Jeane Dixon's newspaper columns is meant only as a very loose guide. To make the advice more precise, the astrologer must look at an individual's birth chart. we have seen that even this more personally tailored advice must remain nonspecific, so it seems that no gain in restriction of interpretations is forthcoming from individualized treatment. At another level, the astrologer seeks support, not for the generalized advice, but for the entire speculative enterprise of astrology, at which point the Rescher strategy can be invoked to show that no accuracy of prediction or success in control is forthcoming from astrology. While this may be true, it is important, I think, to look briefly at what at least some astrologers present as support for their activity.

In the absence of an account of how positions of heavenly bodies can be causally related to either specific or general advice, astrologers look for support for their speculations elsewhere. A common move amounts to an appeal to authority: Aquinas's acceptance of astrology (so long as necromancy is not included) as a science, and Kepler's casting of astrological charts. Parker and Parker appeal to Jung's interest in astrology as "the first important stage in the slow revival of scientific interest in the topic [the causal relationship between planetary movements and events in human life] which we are witnessing today." They continue, "What is important ... is the fact that a man recognized throughout the world as one of the greatest and most provocative thinkers of the century should have considered the data of astrology sufficiently interesting in principle not only to include it as an integral part of his own cosmological theory."[32]

If we drop the vacuous appeal to authority, what is left to support

---

[32] D. Parker and J. Parker, *The Compleat Astrologer*, p. 59.

astrology and provide warrant for limiting interpretations of its claims? I think the obvious answer is "nothing." On the question of causal influence of planetary motions on human action, some astrologers point out that various forms of action at a distance are allowed by scientists, that the biological clock is an influence akin to astrological causality, and the like (ibid., 50-57). Such appeals simply complicate the speculations, lending unsubstantiated additional claims to the already heavy baggage of interpretation involved in astrology. As if this were not enough, astrologers often seek to find support for their speculations by appealing to the long history of widespread interest in their "science," and to such seemingly unrelated and highly symbolic matters as the pyramids, Stonehenge as an ancient cosmic computer, the celestial bestiary, the fact that traditionally signs of the Zodiac have been related to various parts of the human body, the close connections between mythology and astrology, even to Culpeper's herbal remedies! (For details, see Parker and Parker.)

It thus appears that, in order to take the texts of astrology to be well supported, we are driven further and further away from considerations of empirical adequacy, and closer and closer to the need for acceptance of a complete, and thoroughly questionable, metaphysical system. That system is built up of the kinds of appeals just listed, all of which boils down to being appeals based on tradition, history, community of interests, authority, collection of apparently meaningfully connected symbols (perhaps even Jung's archetypes of the collective unconscious). The consequence is that astrology cannot have its texts restricted to 1 or a few interpretations. Indeed, astrology appears as the pseudoscience *par excellence* because it is the "science" of interpretation *par excellence*. Since the range of permissible hermeneutic insight required to understand astrological claims is so large, the kind of support it seeks cannot place it close to the sciences, whose texts have a narrow range of permissible interpretations that approaches 1.

Creationism and Its Support

The same is true of creationism, although for very different reasons. Astrology simply requires acceptance of too much nonscience in order to be considered a science. the difficulty with creationism, as an alternative to scientific biology, especially the theory of evolution, is one of a multiplicity of permissible *translations*. What I will now put forward is not new, but I hope the new context I am endeavoring to supply will justify saying it once more. In brief, creationism accepts the account of the creation of life forms found in the book of Genesis of the Bible. God created everything in the heavens and on earth in six days, starting at a time. Creationists disagree about the year in which God began his work (as well they might; it is difficult to imagine how years could be counted *before* God created the heavens!). Because creationists believe this account is a true theory, they of course want it to be taught in biology classes, as a theory in competition with the theory of evolution.

If the book of Genesis is literally construed, how are we to understand the support for its claims? At the end of the first day of creation, there were God, and light and darkness. How long was that day? We are not told how God measured time. We cannot proceed with the attempt to confirm the claim that on the first day God made light and darkness. furthermore, God's spirit was moving on the waters when he created light and darkness. Are the waters created or not? Presumably not. Again, an incoherent claim. Obviously, literal construal of the book of Genesis, or of any book of the Bible, is impossible. We could at this point simply conclude that the text of the Bible is literary, not literal, and hence is of course subject to differing interpretations. But this would not be fair to the creationists, who are also fundamentalists, and who want to teach Genesis alongside Darwin because they think that what Genesis says is not just to be literally construed, but is literally true.

The problem for the creationists lies elsewhere. To which book of Genesis do we appeal? The original Hebrew version? Luther's translation? Numerous versions of the Bible are available: the 1611 King James version, the 1979-1982 new King James version, the Revised Standard version, the New Revised Standard version (noted for its attribution of gender-free language to God), the New English Bible version, the Revised English Bible (in an ecumenical effort, "Let us make man in our image" becomes "Let us make human beings in our image" – creating a hermaphroditic deity, now we are getting somewhere!), the New International version (the best-selling version in the United States), the New Jewish version, the new American Bible (the most widely recognized Roman Catholic version) and more. We have seen that in the case of scientific languages literally construed, translation from one language into another is relatively unproblematic because of the stability of meaning of technical terms and the absence of metaphor. This does not make scientific texts fun to read, but surely makes them easy to translate.

So we see that the problem for the creationist is twofold, if the desired limitation of the range of permissible interpretations is to be accomplished. First, it must be shown that the text of Genesis can be literally construed, so that it is at least in principle capable of empirical adequacy. As suggested, a literal construal of the text of Genesis turns its sentences into either falsehoods or logical incoherencies. Second, it must be shown either that one version of the Bible is the only correct one, or that all versions are intertranslatable without loss of logical form or of literal reference. How can either alternative be rationally sustained? It cannot.

There is a final move open to the creationist. For it is possible to argue that there are sciences and sciences, and that creationism is a human, not a natural, science. The support for construing any version of the Bible literally comes not from the stability of reference of technical terms and from the absence of metaphor, but from facts having to do with the actions of human beings – again, from history, community, tradition, authority, and shared faith. Restriction of the range of hermeneutic insight required for interpreting biblical texts comes not from the collection of more confirming instances (corroboration), but from the conversion of more witnesses providing sympathetic testimony (vindication). If one seeks to justify the scientific status of creationism along this line, it would appear that creationism, like the human sciences generally, needs to provide adequate justification for the new kinds of evidential appeal. It cannot make good its claims to equal treatment on natural scientific grounds, but must appeal to what until further notice must be regarded as nonscientific forms of evidence.

I hope to have shown that at least two systems of speculation generally thought to be pseudosciences are indeed pseudosciences, and that the reason for this is that their ranges of permissible interpretations are either large, or can only be restricted by appealing to kinds of support which are scientifically unavailing. On my scale of ranges of permissible interpretations, where do the other pseudosciences Laudan lists find themselves? None of them have ranges that approach 1. Some – Rosicrucians, the-world-is-about-to-enders, probably charioteers of the gods – have ranges at the other limit; their texts are gibberish. Some may even fall off the scale and approach insanity: flat earthers, circle squarers, *perpetuum mobile* builders. Still others, literally construed, can be either confirmed or disconfirmed. Nessie might one day be found and captured, Big foot might one day be bagged. To date, alleged sightings of both creatures seem doubtful. So-called anomalous phenomena like the two monsters and UFOs may be worthy of scientific investigation, if we think society can bear the cost. Primal screamers have only themselves to blame.

CHAPTER 13

THE RECEPTION OF GERMAN SCIENTIFIC PHILOSOPHY IN NORTH AMERICA: 1930-1962[1]

It is quite widely believed that something like a consensus in North American philosophy of science[2] existed in the period roughly covering 1930 to 1960. I choose 1930 because it dates the beginning of publication of the journal *Erkenntnis* (edited initially by Reichenbach and Carnap), and opens a decade of something like editorial control over major work in philosophy of science and related fields. Think in this connection of the two series, "*Schriften zur Wissenschaftlichen Weltauffassung*" (edited by Philipp Frank and Moritz Schlick), and "*Einheitswissenshaft*" (edited initially by Otto Neurath, Carnap, Frank, and Hans Hahn). The first of a number of major conferences in analytic philosophy of science was held in Paris in 1935. In 1938 the publication of the *International Encyclopedia of Unified Science* began, under the initiative of Neurath, with Carnap and Charles Morris as associate editors. I note with interest that the

---

1 [Previously published in Maria Carla Galavotti and Alessandro Pagnini (eds.), *Experience, Reality, and Scientific Explanation: Essays in Honor of Merrilee and Wesley Salmon* (Dordrecht: Kluwer, 1999), pp. 1-14.]

2 Although I will use the phrase 'philosophy of science', it is not to be understood in any of its more recent meanings, whatever they may be. The philosophers of science I will be discussing all subscribed to the *scientific philosophy*, in something like the sense given to that position in Hans Reichenbach's *The Rise of Scientific Philosophy*. The scientific philosophy stands in opposition to other philosophies, whether or not they are philosophies of science. Here is a brief statement of the character of scientific philosophy from Reichenbach's book (p. 308):

> There is a body of philosophical knowledge. Philosophy is no longer the story of men who attempted in vain to "say the unsayable" in pictures or verbose constructions of pseudological form. Philosophy is logical analysis of all forms of human thought; what it has to say can be stated in comprehensible terms, and there is nothing "unsayable" to which it has to capitulate. Philosophy is scientific in its method; it gathers results accessible to demonstration and assented to by those who are sufficiently trained in logic and science. If it still includes unsolved problems subject to controversy, there is good hope that they will be solved by the same methods as those which, for other problems, have led to solutions commonly accepted today.

founding editor of *Philosophy of Science*, William Malisoff, was also a member of the Advisory Committee of the *Encyclopedia*. *Philosophy of Science* began publication in January 1934; the lead article, "On the Character of Philosophic Problems", is Malisoff's translation of Carnap's essay originally written in German. All of the developments just outlined took place within what we now regard as the discipline of analytic philosophy. The tool of philosophy is the new and more powerful logic of Russell and Whitehead, Frege, Boole and DeMorgan. The romantic philosophies of the past and present are at best instances of poetry to be appreciated aesthetically, at worst cases of dysfunctional fantasizing.

I choose 1962 as the date of termination of the consensus within logical empiricism because it marks the date of publication of Kuhn's *Structure of Scientific Revolutions*, a work which, interestingly, first appeared in the *International Encyclopedia of Unified Science*.[3] The writings of Paul Feyerabend became influential in the 1960s. The received view was not helped by the fact that it had now (for example, in the writings of Israel Scheffler, a former student of Nelson Goodman) to combat the perceived *subjectivism* of Kuhn and Feyerabend.[4]

Much that was characteristic of scientific philosophy in America owed its origins to the arrival of German-speaking philosophers in the United States in the 1930s and 40s. Feigl left Austria in 1930, Reichenbach, Cassirer and Paul Schrecker left Germany in 1933, Carnap in 1935, all of whom eventually finished their careers in U. S. universities. This is only to name some of those who escaped the tyranny of Nazism. Fortunately there were others among them, including natural and social scientists. They left a profound philosophical legacy, one that reshaped not only philosophy in America, but also in the rest of the world.

Permit me a personal recollection at this point. As an Instructor in philosophy in the University of Pennsylvania during 1952-54, I taught a two-semester course in logic and philosophy of science. The text for the first semester was Tarski's *Introduction to Logic*. In 1952-53 in the second semester we used the blue pamphlets of the *Encyclopedia*: Carnap (logic and mathematics), Morris (theory of signs), Nagel (probability), Hempel (concept formation), and Woodger (theory construction). The following year the Feigl and Brodbeck *Readings* appeared, and we used selections from that anthology. I have sometimes pondered the thought: suppose Kuhn's work had been available in the blue pamphlet format – what kind of difference would that have made to my education in orthodoxy?

But Kuhn, whose book was soon to be regarded as holding ideas in opposition to those of logical empiricism, was not available, and those of us forced

---

3 Neurath's introduction to this work distinguishes between the cooperative empirical synthesis that results in the never-completed encyclopedia, and that "metaphysical comprehensiveness" that results in completed idiosyncratic systems.

4 Scheffler's critique, successful or not, introduced an unfortunate understanding of these philosophers. Carnap wrote an accepting review of *The Structure of Scientific Revolutions*. From the beginning, those who worked on the *Encyclopedia* counted Kuhn as one of their own. For more on Scheffler see note 9 below.

by circumstances to accept without question the distinction between the context of justification and the context of discovery could not have dreamed in our wildest moments that Feyerabend would burst onto the scene a decade hence. When we were not worrying over details of Nelson Goodman's *The Structure of Appearance*, helping Richard Martin (who had been a student of Carnap at Chicago) to read proofs of his work in formal semantics, agonizing over the meaning of analyticity in Henry Hiz's seminars,[5] we had only Paul Schrecker's provision of the fascinating scientific background of the thought of Leibniz and Kant to console us. But some of us were released from the isolation ward early, and I, for one, found the consolation of Schrecker's Leibniz and Kant more than enough reward for having caught the logical empiricist disease. End of personal recollection.

What happened in this three-decade period to achieve something like consensus in philosophy of science? Why did it come about? Since I was trained by those who inherited the orthodoxy, let me begin by saying what I think the planks in the empiricist platform I was recruited to stump for are. The article by Carnap alluded to above sets the tone.[6]

> In order to discover the correct standpoint of the philosopher, which differs from that of the empirical investigator, we must not penetrate *behind* the objects of empirical science into presumably some kind of transcendent level; on the contrary we must take a *step back* and *take science itself as the object. Philosophy is the theory of science...Philosophy is the logic of science.*

Carnap goes on to tell us that we must carefully distinguish philosophy, as analysis of the logic of science, from psychology and sociology of science, because psychology and sociology are empirical sciences. We thus have the following meaty claims:

1. Philosophy is distinct from empirical science.

2. Philosophy of science (indeed *philosophy itself*) is analysis of the logic of science.

3. Proper philosophy of science (*philosophy itself*) is anti-metaphysical.

4. [By implication] The only proper philosophy is philosophy of science.

5. [By implication] The context of justification, as part of the logic of

---

5 Quine had demonstrated in "Two Dogmas of Empiricism", read by us with almost religious awe, that all was not well with the analytic/synthetic distinction. It was a sign of the times that one could not get a Ph.D. in philosophy from a first-rate American university without producing one's own explication of the notion of analyticity.

6 Compare Carnap's characterization of philosophy with that of Reichenbach quoted in note 2 above.

science (epistemology is absorbed into philosophy of science), is distinct from the context of discovery as part of the empirical study of science in psychology and sociology (and perhaps also history, although it is interesting that Carnap does not mention history as included among the empirical sciences[7]).

What does this view of the character of philosophy of science yield by way of a theory of science? The major features of the theory are well known. I will only briefly list those features I take to be most definitive of the logical empiricist theory of science.

1. A scientific theory is a set of sentences of differing degrees of generality.

2. In the best theories, most of the ingredient sentences are deductively connected.

3. Some of the generalizations of the theory are laws.

4. Such a theory must be empirically testable, a condition requiring that some sentences in the theory are linked to a theory-neutral observation language.

5. The observation language makes possible experimental laws, sentences deductively entailing statements describing the world. In the best theories, these descriptive statements also turn out to be deductive consequences of the theoretical laws of the theory.

6. Theoretical terms have a different epistemic status than observational terms.

7. The aim of a theory is to establish explanatory laws.

8. Explanation by means of such theories is hypothetico-deductive, and all

---

[7] Parallel developments in the late 19th and early 20th centuries (especially dependent upon Hegel's idealism) displayed the absorption of philosophy into the history of philosophy. Carnap, Reichenbach, and perhaps most especially my teacher, Nelson Goodman, had no patience with those who worked in the history of philosophy. These orthodox logical empiricists would not have understood that history and philosophy of science or *Wissenschaftstheorie* are perfectly legitimate academic subject matters. But of course it was Kuhn's theory of the history of science and Feyerabend's ideas about the dialectical development of science that were to spearhead the new developments that replaced the sterility of a too-restrictive logical analysis of science with a more robust kind of understanding of science. It is to be regretted, however, that for many Kuhn and Feyerabend came to symbolize the utter subjectivity of all cognitive decisions, and the relativity of social (including academic and scientific) communities. Their work deserved a better reward.

that remains of the concept of causality is absorbed into this form of explanation.

9. Science is cumulative and progressive: the theories become ever more general as science progresses.

10. Science is reductive: less general laws are 'reduced' by being shown to be consequences of more general laws, providing also reason to think that the 'entities' talked about in the less general theory are 'really' the kind of entity talked about in the theory producing the reduction.

One recognizes here elements of the positivist legacy. It would be a mistake, however, to think that the ten points just listed received monolithic support from all those who were associated with logical empiricism. Just consider the case of Hans Reichenbach, who vigorously opposed positivism (especially the phenomenalism so typical of the thought of the early positivists) and the rigidity of the hypothetico-deductive reconstruction of the logic of science, replacing it with his own, subsequently widely accepted, probabilistic interpretation. Or of Carnap, whose principle of tolerance introduces the idea that the choice of logics is a matter of preference, and who argues in the *Aufbau* that there is no significant epistemological difference between his own constitution system and realism, transcendental idealism and phenomenalism. It is all a matter of choice of a linguistic framework. Carnap's conventionalist leanings are absorbed by Goodman into *The Structure of Appearance*, and eventually emerge as ways of worldmaking. Carnap's linguistic framework is the ancestor of Goodman's worlds, which are systems of symbols. Or of Hempel, whose thought evolved into more complex patterns as he dealt with one after another of the problems *within* logical empiricism. Or of Frank, who rejected the concept of science as cumulative.[8]

---

[8] In 1954, Philipp Frank gave a lecture for the American Academy of Arts and Sciences in Boston. In the audience, seated side-by-side, were Tom Kuhn and Adolf Grünbaum. Frank was discussing change in science, progress in science. He stated that if one really wants to understand scientific progress, one should consider the example of a woman buying a dress. (These were more generous times-political correctness meant something quite different in 1954 than it does now. Senator McCarthy was finding communists everywhere.) The woman will look for a dress that fits properly the changes in her body that have taken place since the last dress was purchased, hopefully at the same time not distorting the good fit of other parts of her body. Thus there are two primary considerations: the dress must fit better than older ones in places that matter, and there must be a better overall fit. Of course some good fit might be lost in attempting to satisfy the second consideration. So also in science we look for satisfaction of the two considerations, now applied to theories. Often a newly accepted theory has lost some of the explanatory power of its earlier rival. Examples are plentiful. The loss of explanatory power of a theory is now widely referred to as 'Kuhn loss'. Years later, shortly before his death, Adolf reminded Tom of Frank's remarks. Tom Kuhn was shocked. It is Adolf's view that Tom had absorbed the lesson offered by Frank, had repressed conscious memory of it, and had, by some mental trickery, called it into consciousness as his own idea. But now we know the truth: one of the fathers of the scientific philosophy actually wrote Kuhn's *The Structure of Scientific Revolutions*.

When the revolution of the 1960s erupted, every one of the ten tenets was severely questioned.[9] The old orthodoxy, of course, was itself once a revolution, and once the sentential theory of theories was introduced, much of the work in philosophy was what followers of Kuhn might call work in 'normal' philosophy: how to make the analytic/synthetic distinction precise? how to resolve the paradoxes of confirmation? how to account for lawlikeness given the limitations of an extensionalist logic and the resulting paucity of modal distinctions? how to incorporate statistical explanation into the H-D model? how to formulate bridge-laws or correspondence rules? how to formulate the empiricist criterion of meaning? And so on. Critics of the orthodox view in philosophy of science were later to point out that each one of these problems is generated, not by an attempt to understand some formulated feature of past or present science, but by the sentential theory of theories itself.[10]

This may be true: Theories of all kinds do, after all, have to keep their own acts clean, and doing the housework can become an end in itself. I personally believe that it is very wrong to blame those who worked hard on details of the orthodox theory for having turned philosophy of science in upon itself. The sentential theory of theories is a respectable one, with good credentials. It is a plausible reconstruction of an empiricist understanding of science that emerged with full force in the eighteenth century and was widely defended in various forms in the nineteenth century. Its advocates were also honest workers: In formulating their problems in new and more precise ways, they themselves found internal difficulties in their own program. The sentential theory may be wrong, but it cannot

---

[9] Here is Scheffler's statement of the outcome of Kuhn's ideas in *Structure*:

> But now see how far we have come from the standard view [what I have been calling the orthodox or received view]. Independent and public controls are no more, communication has failed, the common universe of things is a delusion, reality itself is made by the scientist rather than discovered by him. In place of a community of rational men following objective procedures in the pursuit of truth, we have a set of isolated monads, within each one of which belief forms without systematic constraints.
>
> I cannot myself, believe that this bleak picture, representing an extravagant idealism, is true. In fact, it seems to me a *reductio ad absurdum* of the reasonings from which it flows.

All that was needed was that professors of deconstruction came to accept what Scheffler rejected. Both philosophy of science and science itself were served with the notice that they had become redundant.

[10] It is important to reflect on the fact that although the founders of the orthodoxy were trained, in the first instance, as scientists, and spent most of their early professional lives in the environment of scientists, when they came to the United States they were immediately absorbed into departments of philosophy. As time went by, more and more *philosophers* of science had little or no training in science.

be put aside without a pretty total revamping of our ways of viewing science.[11]

It is not my intention in this paper to show how this orthodoxy was brought to ruin (if indeed it has been).[12] Rather, I want to aid understanding of how this orthodox theory of scientific theories came about. After all, it is no small thing for a certain view of science to hold sway for three decades (discounting for the moment its historical roots in Hume and others), and understanding why this happened could help us to understand many other related matters; for example, it could help us to understand those forces in intellectual history that begin in different areas and then converge to produce what might be called "the philosophy of the era." The hard work of producing an account of the origins of the scientific philosophy of the years 1930-1962 has only just begun, and I think we may be in for some surprises. For now, I can only offer a sketch of an account I would be willing to try to give. At this stage, the sketch is best viewed as a set of tentative speculative hypotheses made plausible by circumstances; even so, it may provide a context for good discussion.

Philosophy of science as we know it today, as an investigation of the nature and the epistemological status of science, began in the nineteenth century with the work of John Herschel, William Whewell, John Stuart Mill, August Comte, Ernst Mach, Pierre Duhem, Helmholtz, Charles Sanders Peirce and a few others. We are only beginning to understand the nineteenth-century preoccupation with *justifying science*. By the middle of that century physical science and its technological applications were a resounding success. Mill and Comte took this to mean that we have now at our disposal *laws of nature*, empirical regularities so firmly established that we can at last expect accurate and precise predictions of future events. The program of Francis Bacon has at last been fulfilled; we can develop a *logic* of induction.

Not all shared the positivism of Mill and Comte. Whewell did not dismiss metaphysics, and held that certain inscrutable mental imputations were involved in scientific induction. Mach admired Whewell's theory of induction, as did Peirce.[13] (I think this is important: The founder of American pragmatism [Peirce] and a major instrumentalist [Mach] both adopted elements of an essentially psychologistic theory of induction, put forward by the most staunch defender of hypothetico-deductivism in the nineteenth century [Whewell].) Despite differences,

---

11 Again Carl Hempel comes to mind as the best representative of the intellectual integrity of logical empiricists. Just recall his work on changes in the empiricist meaning criterion, and his more recent efforts to modify the H-D model of explanation and to locate criteria of rationality at work in the activities of scientists.

12 There are signs of a modest revival, resulting from a more sensitive recognition of the merits of early logical positivism. The work of Michael Friedman leads the way.

13 There is evidence that Maxwell's work in science consciously employed elements of Whewell's methodology; Darwin's copy of Whewell's *Philosophy of the Inductive Sciences* is richly annotated.

I think that what eventually issued from the theorizing about science of these philosophers and scientists is a kind of *epistemological scientism*, a kind of commitment to the view that science will cure all of our epistemological ills. Scientism linked to successful science; after Herschel and Whewell, Maxwell and the Cavendish Laboratories; after Mach, relativity physics. And scientism linked finally to philosophical schools: the Vienna Circle, the Berlin group, the Warsaw group; and to movements: pragmatism, early twentieth-century American realism.

Please consider: By the turn of the century, what were the philosophically respectable alternatives to this epistemological scientism, *so far as understanding science is concerned*? Dilthey and Husserl and the move to hermeneutics might have provided one alternative, but it has not been until quite recently that we have had promising, if as yet inadequate, formulations of this approach in *philosophy of science*, for example, in the work of Patrick Heelan. Philosophical idealisms of various kinds might have provided a second alternative, both those stemming from Hegel and that of Bergson. But the idealism that prevailed in England, Germany, America and France was essentially anti-science in orientation.[14] For Bergson scientific knowing is debased knowing; only direct intellectual sympathy can give us reliable knowledge of things. For the Hegelians, history supplants nature; physical science is an epiphenomenon of logic in motion, of *Geist*. (I am prepared to hazard the hypothesis that the large number of departments of philosophy *and* religion that grew up in U. S. colleges and universities in the United States were responding favorably to the dominant idealism and its careless anti-scientific outlook. Philosophy is to keep science at arm's length, and the best way to achieve the distance is to ally philosophy with religion.)

I suggest that in fact there was no proper alternative to the epistemological scientism that represented the early prefiguring of the sentential theory of theories, and that (again, in fact) the best nineteenth-century views of science originated in close connection with actual developments in physical and formal science. There is perhaps an irony here. We are now told that the orthodoxy of 1930-1962 paid little attention to 'real' science. The fact is that in its prototypical form this philosophy of science developed out of an understanding of current science in the making. Herschel, Mach, Helmholtz and Whewell were active scientists. Whewell and Duhem – ahead of their time – both insisted that proper philosophy of science can only come from a close study of the history of science. It was also not clear who was the scientist, who the philosopher. Einstein and Whitehead were members of the first editorial board of the journal *Philosophy of Science*; Russell and Dewey served on the advisory committee of the *Encyclopedia*.

The current science that was then in the making included most importantly relativity physics and the formal logic of *Principia Mathematica*. The scientific origins of the new orthodoxy are clearly Continental and British. But the growing

---

[14] Recall that Scheffler refers to Kuhn's view as an 'extravagant idealism'. The old dialectical clashes may become subdued; they never completely disappear.

threat of the Second World War moved most of the remaining players to the United States and Great Britain.[15] In Britain Richard Braithwaite represented the new orthodoxy (and did it very well indeed). Most of the action, however, took place in the United States; I will limit the remainder of my discussion to developments there.[16]

My next historical thesis is: *By the time logical empiricism moved to the United States, the welcoming soil had already been well prepared* I have indicated that much of American philosophy in the early part of this century was decisively anti-scientific; it provided no alternative within philosophy of science to the new scientism. However, philosophy of science itself had already begun to distinguish itself as a scientific speciality in some departments of philosophy in the United States. Think of the advent of pragmatism at Harvard and the highly significant influence of C.I. Lewis, the work of F.S.C. Northrop at Yale (and the visits of Cassirer and Whitehead), John Dewey's unrelenting defense of the 'scientific attitude,' the attitude of dispassionate, objective consultation of nature, and the little known but very influential work of Edgar Singer at University of Pennsylvania, work that would shape the ideas of C. West Churchman, Richard Rudner, and the present author.

We sometimes fail to appreciate how decisive some of these growing inter-linkages became. In effect, Dewey was recommending that the future of philosophy was not as a separate subject matter, but as a collection of problems to be solved by the social scientists. The model of a student of Dewey is the person who abandons philosophy for psychology or anthropology. In a similar vein, Singer promoted the connection between hard-edged pragmatism and operations research, between the economy of thought and the economy. Again, we see these philosophers at work in or close to the new pragmatically based sciences whose further development they championed. In this climate the journal *Philosophy of Science* was born at the University of Pennsylvania.[17]

---

15 Including Ernst Cassirer, who with his fellow neo-Kantians in Marburg was the first to recognize that Kant's *Kritik der reinen Vernunft* is a work in philosophy of science. We have only recently begun to appreciate how powerful the influence of that book has been on later scientists and philosophers of science.

16 I am deliberately eliminating Popperianism from consideration. Popper's historical position is ambiguous. His falsificationism produced a legion of loyal followers, most of them opposed to the views of the major orthodox figures. On the other hand, Popper is the strongest advocate of the view that the logic of science is hypothetico-deductive, and from this point of view can perhaps be included amongst the orthodox. Only 'perhaps'. His insistence that the logic of science is exclusively deductive, and that therefore there is no induction, does not sit well with the prevailing orthodoxy.

17 I cannot resist the personal note that after graduation from the University of Pennsylvania my first offer of employment came, not from a college or university, but from the Burroughs Adding Machine Company in Philadelphia. The post they envisaged would not have involved me in the mathematics of computers, a move that would have delayed development of the great Philadelphia mainframe almost indefinitely, but was one in which I would have spent my time trying to persuade the U. S. military of the great promise of computers. In effect, I would have been propagandizing on behalf

The fertility of the soil that nurtured orthodoxy was not engendered by pragmatism alone. It was guaranteed as well by those philosophers who called themselves "realists," "neorealists," and "critical realists," all of them arch enemies of idealism. For these philosophers (among them: R. B. Perry, Sellars *pére*, Holt, Spading) philosophical realism appealed to scientific realism as a central part of its data base. Consider the following anti-idealist planks in "Program and First Platform of Six Realists"[18] (*Journal of Philosophy* 1910):

> 1. The entities (objects, facts, etc.) under study in logic, mathematics and the physical sciences are not mental in any usual or proper meaning of the word "mental."
>
> 2. The being and nature of these entities are in no sense conditioned by their being known.
>
> 17. Realism, while admitting the tautology that every entity which is known is in relation to knowing or experience or consciousness, holds that this knowing, etc., is eliminable, so that the entity is known as it would be if the knowing were not taking place. That is to say, the entity, in its being, behavior, and character, is independent of the knowing. This position agrees with common sense and with science in holding (1) that not all entities are mental, conscious, or spiritual, and (2) that entities are knowable without being known.

Apart from the fact that it is interesting to see that for these philosophers scientific realism is crucially required as part of the argument against idealism, it is also important to see how much emphasis is here placed on the contention that, construed as theoretically refined and extended common sense, successful science is itself the primary datum of the philosophical realism. I do not think we can overestimate the support American philosophical realism gave to the arrival of the new scientism. Just consider the line from Sellars *pére* to Sellars *fils*: from "If idealism involves solipsism, mental pluralism of the empirical sort which admits communication and mutual knowledge must involve realism," to "[discussion of realism is] one of the most significant dialogues of our time."

There are ironies here, too. It may seem inappropriate that an articulate position in ontology should be appealed to as part of the support staff for an anti-

of the creation of what is now the Internet. Dick Rudner's entry into academe was delayed for some years while he worked as an operations research consultant to the United State Navy and the Pennsylvania Railroad. Two philosophers later to become editors of *Philosophy of Science* were thus initially thought of as specialists in the pragmatics of persuasion and the technicalities of proper placement of personnel and rolling stock.

18 'Platform,' in its use in ordinary English, is a list of political or ideological commitments. For these despisers of idealism, there is no sharp line that divides ideology and philosophy. On all fronts, the victory for naturalism is thought to be a forgone conclusion. It is to be noted that *seventeen* philosophers signed themselves as authors of this platform. Such mutual agreement is surely rare in philosophy. In politics, however, one counts votes.

ontological philosophy of science. Recall Carnap's important distinction between internal and external ontological questions, and the privileged status given to nominalism amongst the faithful. Nelson Goodman issues the challenge:

> You may decry some of these [nominalist] scruples and protest that there are more things in heaven and earth than are dreamt of in my philosophy. I am concerned, rather, that there should not be more things dreamt of in my philosophy than there are in heaven and earth.

To be sure, discussion of the problems of realism does not finally get off the ground until orthodoxy has been eliminated, and ontology seems once again to have been made respectable. This fact is not incompatible with the claim that American realists, in effect, provided some of the philosophical propaganda on behalf of the new scientism. In the United States, at least, pragmatism, realism, and positive science go hand-in-hand. And in the United States, at least, speculative philosophy, idealism, and rejection of the epistemic credentials of science also go hand-in-hand.[19]

Let me repeat the historical thesis here under investigation: Logical empiricism, as the heir of nineteenth-century epistemological scientism,[20] once transported to the United States, found itself a ready home. Philosophers of science in the United States, as in the early days in Europe, were working close to the leading edge of certain sciences, the triumphs of relativity physics and *Principia Mathematica* were incorporated without important argument into the mainstream of thought about science, and the last vestiges of a philosophy shaped by the desire for an other-worldly metaphysics were decisively eliminated from professional philosophy of science. What could now be produced without difficulty was a philosophy of science free of ontology, of psychology, and of history. The soil was ready to grow both this philosophy of science freed from other parts of what was once assumed to be philosophy, and, as Reichenbach put it, the first genuine *scientific philosophy*.

What I am suggesting is that an extraordinary confluence of philosophically compatible commitments took place in the United States in the

---

[19] In an article anthologized in one of the Biblical texts of logical empiricism – *Readings in Philosophical Analysis* – Herbert Feigl refers to William James's distinction between the tough-minded and the tender-minded, listing among the former empiricists, skeptics, naturalists, positivists and pragmatists, and among the latter, speculative metaphysicians, intuitionists, rationalists and absolute idealists. Would that it were all that simple! It is to be regretted that Feigl and other tough-minded empiricists did not live long enough to witness Nicholas Rescher's attempt systematically to consolidate the gains made by pragmatism, realism, and idealism, all in the context of *acceptance* of the epistemic priority of science, and all worked out with scrupulous attention to the canonical demands of good analytic philosophy.

[20] I apologize for use of this term, which in normal usage, has a pejorative sense: Science can answer all of our questions – out with religion, out with metaphysics, out with all forms of irrationality. What I intend the word to denote, in this context, is natural science as the best-confirmed source of reliable knowledge.

interval between the two world wars. It may be, indeed, that the advent of the second World War made any retreat from the established orthodoxy impossible, at least until the next generation of philosophers had been trained and were in place. Certainly in the late 1940s and 1950s there were again no serious alternatives to logical empiricism as philosophies of science. In the United States and the United Kingdom the task of rebuilding the world held nothing in common with the despair and anxiety of the existentialist response.[21] The choices were not choices between competing philosophies of science, but between a philosophy that is anti-science and one that is pro-science, and if one chose the latter, one was choosing logical empiricism and the sentential theory of theories. The theory of science of dialectical materialism was not yet to be heard from in a serious way; as I have mentioned, the hermeneutic response would not become serious until roughly twenty years after Kuhn's revolutionary treatment of scientific revolutions. By that time, we also had the anti-realism championed by many, the irrealism of Goodman, and the semantical theory of scientific theories of Bas van Fraassen; we also had history and philosophy of science (*Wissenschaftstheorie*) as a distinct and well-recognized discipline. We have now also to face the growing influence of sociological studies of scientists at work: why pay attention to historical figures in the history of science and philosophy when we have living scientists to study?

Logical empiricism may not be dead; it has simply been enjoying a rejuvenating nap. As Gereon Wolters has persuasively argued, the nap has already been interrupted by those engaged in serious analytic investigation of philosophical problems generated by biology. Furthermore, the creation of logical empiricism was motivated by studies in the foundations of logic and mathematics, space-time theories, and the foundations of physics. There have been no important and damaging threats to studies in these areas.

So long as scientists theorize and subject their theories to test, philosophers of science will analyze the structure of the theories and the logic of experimentation. The continuation of some form of analytic logical empiricism seems to me to be guaranteed. But of the future of *the scientific philosophy* one must not expect too much. The irrationalities of deconstructionism are too seductive, too easy. Some will suggest that by now logical empiricism is of merely antiquarian interest, somewhat on the style of the arguments for God's existence in Descartes's *Meditations*. But we must beware of such a conclusion. Paul Feyerabend, believing as he did in the powers of witch doctors, nevertheless took prescribed medication for the chronic pains in his head. In a strict sense the scientific philosophy is a *discipline*, and disciplined thought is not very popular just now. If the scientific philosophy disappears, so does much of the collective sanity of civilized human beings.

---

[21] It is of great interest to note that existentialist-inspired ideas only came much later to North America, and they arrived, not because of matured philosophical interest in 'Continental' philosophy, but because of the efforts, largely anti-scientific in nature, of religious thinkers and literary critics. The 'postmodern' revolt has created conferences proclaiming the death of science, and we are encouraged not to become mourners.

The Western Ontario Series in Philosophy of Science

17. M. Ruse: *Is Science Sexist?* And Other Problems in the Biomedical Sciences. 1981
ISBN 90-277-1249-2; Pb 90-277-1250-6
18. N. Rescher: *Leibniz's Metaphysics of Nature.* A Group of Essays. 1981
ISBN 90-277-1252-2; Pb 277-1253-0
19. L. Laudan: *Science and Hypothesis.* Historical Essays on Scientific Methodology. 1981
ISBN 90-277-1315-4
20. W.R. Shea (ed.): *Nature Mathematized.* Historical and Philosophical Case Studies in Classical Modern Natural Philosophy. 1983 ISBN 90-277-1402-9
21. M. Ruse (ed.): *Nature Animated.* Historical and Philosophical Case Studies in Greek Medicine, 19th-Century and Recent Biology, Psychiatry, and Psycho-analysis. 1983 ISBN 90-277-1403-7
22. W.R. Shea (ed.): *Otto Hahn and the Rise of Nuclear Physics.* 1983 ISBN 90-277-1584-X
23. H.F. Cohen: *Quantifying Music.* The Science of Music at the First Stage of the Scientific Revolution (1580-1650). 1984 ISBN 90-277-1637-4
24. R.E. Butts: *Kant and the Double Government Methodology.* Supersensibility and Method in Kant's Philosophy of Science. 1984 ISBN 90-277-1760-5; Pb 90-277-2384-2
25. J.R. Brown (ed.): *Scientific Rationality.* The Sociological Turn. 1984 ISBN 90-277-1812-1
26. F. Wilson: *Explanation, Causation and Deduction.* 1985 ISBN 90-277-1856-3
27. J.C. Pitt (ed.): *Change and Progress in Modern Science.* 1985 ISBN 90-277-1898-9
28. H.B. Hollinger and M.J. Zenzen: *The Nature of Irreversibility.* A Study of Its Dynamics and Physical Origins. 1985 ISBN 90-277-2080-0
29. K. Okruhlik and J.R. Brown (eds.): *The Natural Philosophy of Leibniz.* 1985
ISBN 90-277-2145-9
30. G. Oddie: *Likeness to Truth.* 1986 ISBN 90-277-2238-2
31. F. Wilson: *Laws and Other Worlds.* A Human Account of Laws and Counterfactuals. 1986
ISBN 277-2232-3
32. J. Earman: *A Primer on Determinism.* 1986 ISBN 90-277-2240-4; Pb 90-277-2241-2
33. R.E. Butts (ed.): *Kant's Philosophy of Physical Science.* Metaphysische Anfangsgründe der Naturwissenschaft (1786-1986). 1986 ISBN 90-277-2309-5; Pb 90-277-2310-9
34. I.B. MacNeill and G.J. Umphrey (eds.): *Advances in the Statistical Sciences.* Festschrift in Honor of Professor V.M. Joshi's 70th Birthday (6 volumes).
Vol. I: Applied Probability, Stochastic Processes, and Sampling Theory. 1987
ISBN 90-277-2393-1
35. I.B. MacNeill and G.J. Umphrey (eds.): *Advances in the Statistical Sciences.*
Vol. II: Foundations of Statistical Inference. 1987 ISBN 90-277-2394-X
36. I.B. MacNeill and G.J. Umphrey (eds.): *Advances in the Statistical Sciences.*
Vol. III: Time Series and Econometric Modelling. 1987 ISBN 90-277-2395-8
37. I.B. MacNeill and G.J. Umphrey (eds.): *Advances in the Statistical Sciences.*
Vol. IV: Stochastic Hydrology. 1987 ISBN 90-277-2396-6
38. I.B. MacNeill and G.J. Umphrey (eds.): *Advances in the Statistical Sciences.*
Vol. V: Biostatistics. 1987 ISBN 90-277-2397-4

The Western Ontario Series in Philosophy of Science

39. I.B. MacNeill and G.J. Umphrey (eds.): *Advances in the Statistical Sciences.* Vol. VI: Actuarial Science. 1987 ISBN 90-277-2398-2
Set ISBN (Vols 34-39) 90-277-2399-0
40. N. Rescher: *Scientific Realism.* A Critical Reappraisal. 1987
ISBN 90-277-2442-3; Pb 90-277-2528-4
41. B. Skyrms and W.L. Harper (eds.): *Causation, Chance, and Credence.* 1988
ISBN 90-277-2633-7
42. W.L. Harper and B. Skyrms (eds.): *Causation in Decision, Belief Change and Statistics.* 1988
ISBN 90-277-2634-5
43. R.S. Woolhouse (ed.): *Metaphysics and Philosophy of Science in the 17th and 18th Centuries.* Essays in Honor of Gerd Buchdahl. 1988 ISBN 90-277-2743-0
44. R.E. Butts and J.R. Brown (eds.): *Constructivism and Science.* Essays in Recent German Philosophy. 1989 ISBN 0-7923-0251-6
45. A.D. Irvine (ed.): *Physicalism in Mathematics.* 1989 ISBN 0-7923-0513-2
46. J. van Cleve and R.E. Frederick (eds.): *The Philosophy of Right and Left.* Incongruent Counterparts and the Nature of Space. 1991 ISBN 0-7923-0844-1
47. F. Wilson: *Empiricism and Darwin's Science.* 1991 ISBN 0-7923-1019-5
48. G.G. Brittan, Jr. (ed.): *Causality, Method and Modality.* Essays in Honor of Jules Vuillemin. 1991 ISBN 0-7923-1045-4
49. W. Spohn, B.C. van Fraassen and B. Skyrms (eds.): *Existence and Explanation.* Essays in Honor of Karel Lambert. 1991 ISBN 0-7923-1252-X
50. J.C. Pitt: *Galileo, Human Knowledge, and the Book of Nature.* Method Replaces Metaphysics. 1992 ISBN 0-7923-1510-3
51. V. Coelho (ed.): *Music and Science in the Age of Galileo.* 1992 ISBN 0-7923-2028-X
52. P. Janich: *Euclid's Heritage: Is Space Three-Dimensional?* 1992 ISBN 0-7923-2025-5
53. M. Carrier: *The Completeness of Scientific Theories.* On the Derivation of Empirical Indicators within a Theoretical Framework: The Case of Physical Geometry. 1994
ISBN 0-7923-2475-7
54. P. Parrini (ed.): *Kant and Contemporary Epistemology.* 1994 ISBN 0-7923-2681-4
55. J. Leplin (ed.): *The Creation of Ideas in Physics.* Studies for a Methodology of Theory Construction. 1995 ISBN 0-7923-3461-2
56. J.E. McGuire: *Tradition and Innovation.* Newton's Metaphysics of Nature. 1995
ISBN 0-7923-3617-8
57. R. Clifton (ed.): *Perspectives on Quantum Reality.* Non-Relativistic, Relativistic, and Field-Theoretic. 1996 ISBN 0-7923-3812-X
58. P.H. Theerman and K. Hunger Parshall (eds.): *Experiencing Nature.* Proceedings of a Conference in Honor of Allen G. Debus. 1997 ISBN 0-7923-4477-4
59. P. Parrini: *Knowledge and Reality.* An Essay in Positive Philosophy. 1998
ISBN 0-7923-4939-3
60. D. Dieks and P.E. Vermaas (eds.): *The Modal Interpretation of Quantum Mechanics.* 1998
ISBN 0-7923-5207-6